Lecture Notes in Mathematics

A collection of informal reports and seminars
Edited by A. Dold, Heidelberg and B. Eckmann, Zürich

137

H. Applegate, M. Barr, B. Day, E. Dubuc, Phreilambud, A. Pultr, R. Street, M. Tierney, S. Swierczkowski

Reports of the Midwest Category Seminar IV

Edited by S. MacLane, University of Chicago

Springer-Verlag
Berlin · Heidelberg · New York 1970

Title No. 3293

TABLE OF CONTENTS

ON CLOSED CATEGORIES OF FUNCTORS

Brian Day

Received November 7, 1969

The purpose of the present paper is to develop in further detail the remarks, concerning the relationship of Kan functor extensions to closed structures on functor categories, made in "Enriched functor categories" [1] §9. It is assumed that the reader is familiar with the basic results of closed category theory, including the representation theorem. Apart from some minor changes mentioned below, the terminology and notation employed are those of [1], [3], and [5].

Terminology

A closed category $\mathcal{V}$ in the sense of Eilenberg and Kelly [3] will be called a <u>normalised</u> closed category, $V\colon \mathcal{V}_o \to \mathcal{E}ns$ being the <u>normalisation</u>. Throughout this paper $\mathcal{V}$ is taken to be a <u>fixed normalised symmetric monoidal closed category</u> ($\mathcal{V}_o$, $\otimes$, I, r, ℓ, a, c, V, [-,-], p) with $\mathcal{V}_o$ admitting all small limits (inverse limits) and colimits (direct limits). It is further supposed that if the limit or colimit in $\mathcal{V}_o$ of a functor (with possibly large domain) exists then a definite choice has been made of it. In short, we place on $\mathcal{V}$ those hypotheses which both allow it to replace the cartesian closed category of (small) sets $\mathcal{E}ns$ as a ground category and are satisfied by most "natural" closed categories.

As in [1], an <u>end in</u> $\mathcal{B}$ of a $\mathcal{V}$-functor $T\colon A^{op}\otimes A \to \mathcal{B}$ is a $\mathcal{V}$-natural family $\alpha_A\colon K \to T(AA)$ of morphisms in $\mathcal{B}_o$ with the property that the family $\mathcal{B}(1,\alpha_A)\colon \mathcal{B}(BK) \to \mathcal{B}(B,T(AA))$ in $\mathcal{V}_o$ is universally V-natural in A for each $B \in \mathcal{B}$; then an end in V turns out to be simply a family α_A: $K \to T(AA)$ of morphisms in V_0 which is universally V-natural in A. The dual concept is called a coend.

From [1] we see that the choice of limits and colimits made in V_0 determines a definite end and coend of each V-functor $T: A^{op} \otimes A \to V$ for which such exist. These are denoted by $s_A: \int_A T(AA) \to T(AA)$ and $s^A: T(AA) \to \int^A T(AA)$ respectively. We can now construct, for each pair A, $\mathcal{B}$ of V-categories with A small, a definite V-category $[A,\mathcal{B}]$ having V-functors S, T, ...: $A \to \mathcal{B}$ as its objects, and having $[A,\mathcal{B}](S,T) = \int_A \mathcal{B}(SA,TA)$. An element $\alpha \in V\int_A \mathcal{B}(SA,TA)$ clearly corresponds, under the projections $Vs_A: V\int_A \mathcal{B}(SA,TA) \to \mathcal{B}_0(SA,TA)$, to a V-natural family of morphisms $\alpha_A: SA \to TA$ in the sense of [3]. It is convenient to call α, rather than the family $\{\alpha_A\}$ of its components, a V-natural transformation from S to T; for then the underlying ordinary category $[A,\mathcal{B}]_0$ is the category of V-functors and V-natural transformations.

Limits and colimits in the functor category $[A,\mathcal{B}]$ will always be computed evaluationwise, so that the choice of limits and colimits made in V fixes a choice in $[A,V]$ for each small V-category A. Included in this rule are the concepts of cotensoring and tensoring,which were seen in [5] to behave like limits and colimits respectively.

In order to replace the category of sets by the given normalised closed category V, we shall "lift" most of the usual terminology. A V-monoidal category $\mathbb{V}$ is a V-category A together with a V-functor $\bar{\otimes}$: $A\otimes A \to A$, an object $\bar{I} \in A$, and V-natural isomorphisms $\bar{a}$: $(A\bar{\otimes}B)\bar{\otimes}C \cong A\bar{\otimes}(B\bar{\otimes}C)$, $\bar{r}$: $A\bar{\otimes}\bar{I} \cong A$, and $\bar{\ell}$: $\bar{I}\bar{\otimes}A \cong A$, satisfying the usual coherence axioms for a monoidal category - namely axioms MC2 and MC3 of [3]. If, furthermore, $-\bar{\otimes}A$ and $A\bar{\otimes}-$: $A \to A$ both have (chosen) right V-adjoints for each $A \in A$, then $\bar{V}$ is called a V-<u>biclosed category</u> (see Lambek [8]). A V-symmetry for a V-monoidal category $(A, \bar{\otimes}, \bar{I}, \bar{r}, \bar{\ell}, \bar{a})$ is a V-natural isomorphism $\bar{c}$: $A\bar{\otimes}B \cong B\bar{\otimes}A$ satisfying the coherence axioms MC6 and MC7 of [3]. Finally we come to the concept of a V-<u>symmetric-monoidal-closed category</u> which can be described simply as a V-biclosed category with a V-symmetry; we do not insist on a "V-normalisation" as part of this structure. An obvious example of such a category is V itself, where $\bar{\otimes}$ is taken to be the V-functor Ten: $V\otimes V \to V$ defined in [3] Theorem III.6.9.

We note here that, for V-functors S: $A \to C$ and T: $B \to D$, the symbol S$\otimes$T may have two distinct meanings. In general it is the canonical V-functor $A\otimes B \to C\otimes D$ which sends the object (ordered pair) $(A,B) \in A\otimes B$ to the object $(SA,TB) \in C\otimes D$, as defined in [3] Proposition III.3.2. When C and D are both V, however, we shall also use S$\otimes$T to denote the composite

$$A\otimes B \xrightarrow{S\otimes T} V\otimes V \xrightarrow{\text{Ten}} V.$$

The context always clearly indicates the meaning.

<u>Henceforth we work entirely over</u> V <u>and suppose that the unqualified words "category", "functor", "natural transformation", etc. mean</u> "V-<u>category</u>", "V-<u>functor</u>", "V-<u>natural transformation", etc.</u>

1. Introduction

Let A be a small category and regard A^{op} as a full subcategory of $[A,V]$, identifying $A \in A^{op}$ with the left represented functor $L^A: A \to V$ in the usual way. For each $S \in [A,V]$ we have the canonical expansion (see [1]) $S \cong \int^A SA \otimes L^A$ which asserts the density (adequacy) of A^{op} in $[A,V]$. If $[A,V]$ has the structure of a biclosed category $\overline{V}$ then, in view of this expansion, the value $S\overline{\otimes}T$ of $\overline{\otimes}: [A,V]\otimes[A,V] \to [A,V]$ at (S,T) is essentially determined by the values $L^A\overline{\otimes}T$, because $-\overline{\otimes}T$ has a right adjoint. These in turn are determined by the values $L^A\overline{\otimes}L^B$, because each $L^A\overline{\otimes}-$ has a right adjoint. Writing $P(ABC)$ for $(L^A\overline{\otimes}L^B)(C)$, we see that the functor $\overline{\otimes}$ is essentially determined by the functor $P: A^{op}\otimes A^{op}\otimes A \to V$, in the same way that the multiplication in a linear algebra is determined by structure constants.

These considerations suggest what is called in section 3 a premonoidal structure on A. This consists of functors $P: A^{op}\otimes A^{op}\otimes A \to V$ and $J: A \to V$, together with certain natural isomorphisms corresponding to associativity, left-identity, and right-identity morphisms, which satisfy suitable axioms; a monoidal structure is a special case. Before attempting to write the axioms down, we collect in section 2 the properties of ends and coends that we shall need.

The main aim of this paper is to show that, from a premonoidal structure on a small category A, there results a canonical biclosed structure on the functor category $[A,V]$; this is done in section 3. As one would expect, biclosed structures on $[A,V]$ correspond bijectively to premonoidal structures on A to within "isomorphism". However we do not formally prove this assertion, which would require the somewhat lengthy introduction of premonoidal functors to make it clear what "isomorphism" was intended.

The concluding sections contain descriptions of some commonly occurring types of premonoidal structure on a (possibly large) category A. The case in which the premonoidal structure is actually monoidal is discussed in section 4. In section 5 we provide the data for a premonoidal structure which arises when the hom-objects of A are comonoids ($\otimes$-coalgebras) in V in a natural way. In both cases the tensor-product and internal-hom formulas given in section 3 for the biclosed structure on $[A,V]$ may be simplified to allow comparison with the corresponding formulas for some well-known examples of closed functor categories.

2. Induced Natural Transformations

Natural transformations, in both the ordinary and extraordinary senses, are treated in [2] and [3]. Our applications of the rules governing their composition with each other (and with functors) are quite straightforward and will not be analysed in detail.

The following dualisable lemmas on induced naturality are expressed in terms of coends.

Lemma 2.1. *Let* $T: A^{op}\otimes A\otimes B \to C$ *be a functor and let* $\alpha_{AB}: T(AAB) \to SB$ *be a coend over* A *for each* $B \in B$. *Then there* exists a unique functor $S: \mathcal{B} \to \mathcal{C}$ making the family α_{AB} natural in B.

Proof. For each pair B, B' $\in \mathcal{B}$ consider the diagram

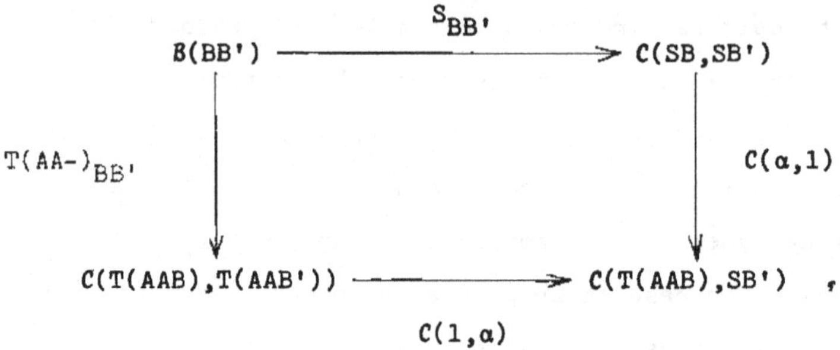

Because $\mathcal{C}(\alpha,1)$ is an end and $\mathcal{C}(1,\alpha).T(AA-)_{BB'}$ is natural in A we can define $S_{BB'}$ to be the unique morphism making this diagram commute. The functor axioms VF1' and VF2' of [3] are easily verified for this definition of S using the fact that $\mathcal{C}(\alpha,1)$ is an end. S is then the unique functor making α_{AB} natural in B.

Lemma 2.2. Let $T: \mathcal{A}^{op} \otimes \mathcal{A} \otimes \mathcal{B} \to \mathcal{C}$ and $S,R: \mathcal{B} \to \mathcal{C}$ be functors, let $\alpha_{AB}: T(AAB) \to SB$ be a coend over A, natural in B, and let $\beta_{AB}: T(AAB) \to RB$ be natural in A and B. Then the induced family $\gamma_B: SB \to RB$ is natural in B.

Proof. For each pair B, B' $\in \mathcal{B}$ consider the diagram

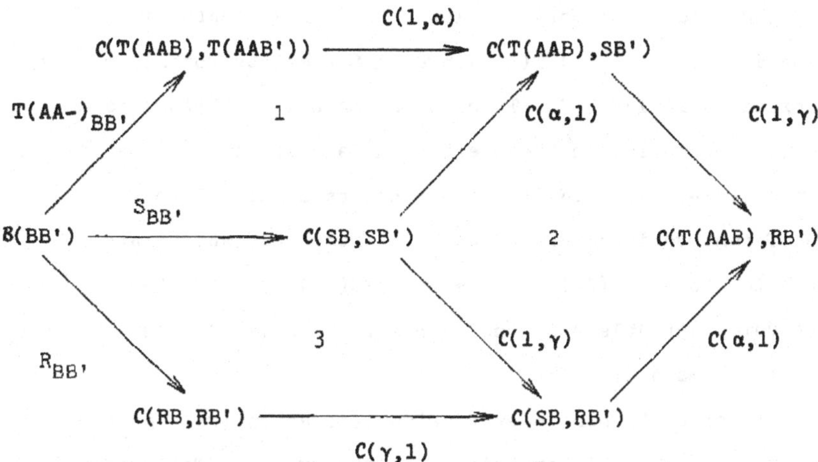

The commutativity of region 1 and that of the exterior express the naturality in B of α and β respectively. Region 2 clearly commutes hence, because $C(\alpha,1)$ is an end, region 3 commutes for each pair $B,B' \in \mathcal{B}$, as required.

By similar arguments we obtain

<u>Lemma 2.3</u>. <u>Let</u> $T\colon A^{op}\otimes A\otimes \mathcal{B}^{op}\otimes \mathcal{B} \to C$ <u>and</u> $S\colon \mathcal{B}^{op}\otimes \mathcal{B} \to C$ <u>be functors, let</u> $\alpha_{ABB'}\colon T(AABB') \to S(BB')$ <u>be a coend over</u> A, <u>natural in</u> B <u>and</u> B', <u>and let</u> $\beta_{AB}\colon T(AABB) \to C$ <u>be natural in</u> A <u>and</u> B. <u>Then the induced family</u> $\gamma_B\colon S(BB) \to C$ <u>is natural in</u> B.

<u>Lemma 2.4</u>. <u>Let</u> $T\colon A^{op}\otimes A \to C$ <u>and</u> $R\colon \mathcal{B}^{op}\otimes \mathcal{B} \to C$ <u>be functors, let</u> $\alpha_A\colon T(AA) \to C$ <u>be a coend over</u> A, <u>and let</u> $\beta_{AB}\colon T(AA) \to R(BB)$ <u>be natural in</u> A <u>and</u> B. <u>Then the induced family</u> $\gamma_B\colon C \to R(BB)$ <u>is natural in</u> B.

Let $\mathcal{A}$ be a category and let $T(AA-)$ be a functor into $\mathcal{V}$ whose coend s^A: $T(AA-) \to \int^A T(AA-)$ over $A \in \mathcal{A}$ exists for all values of the extra variables "-". Then, by Lemma 2.1, $\int^A T(AA-)$ is canonically functorial in these extra variables. In the special case where $T(AA-) = S(A-)\otimes T(A-)$ for functors S and R into $\mathcal{V}$ (with different variances in A) we will frequently abbreviate this notation to s^A: $S(A-)\otimes R(A-) \to S(A-)\underline{\otimes}R(A-)$, leaving the repeated dummy variable A in the expression $S(A-)\underline{\otimes}R(A-)$ to indicate the domain of $\int^{\cdot}$.

In order to handle expressions formed entirely by the repeated use of $\underline{\otimes}$, it is convenient to introduce the following considerations which we do not formalise completely. To each expression $\underline{N}$ which is formed by one or more uses of $\underline{\otimes}$, there corresponds an expression N in which each $\underline{\otimes}$ is replaced by $\otimes$, the dummy variables in $\underline{N}$ becoming repeated variables in N; for example, if $\underline{N}$ is $(RA\underline{\otimes}S(AB))\underline{\otimes}T(BC)$ for functors $R\colon \mathcal{A} \to \mathcal{V}$, $S\colon \mathcal{A}^{op}\otimes\mathcal{B} \to \mathcal{V}$, and $T\colon \mathcal{B}^{op}\otimes\mathcal{C} \to \mathcal{V}$, then N is $(RA\otimes S(AB))\otimes T(BC)$. Moreover, there is a canonical natural transformation $q = q_N\colon N \to \underline{N}$ defined, as follows, by induction on the number of occurrences of $\underline{\otimes}$ in $\underline{N}$. If $\underline{N}$ contains no occurrence of $\underline{\otimes}$ then $N = \underline{N}$ and $q_N = 1$; otherwise $\underline{N} = \underline{N'}\underline{\otimes}\underline{N''}$ and q_N is the composite

$$N'\otimes N'' \xrightarrow{q'\otimes q''} \underline{N'}\otimes\underline{N''} \xrightarrow{s} \underline{N'}\underline{\otimes}\underline{N''}.$$

In the above example, q is the composite

$$(RA\otimes S(AB))\otimes T(BC) \xrightarrow{s\otimes 1} (RA\underline{\otimes}S(AB))\otimes T(BC) \xrightarrow{s} (RA\underline{\otimes}S(AB))\underline{\otimes}T(BC)$$

and this is natural in A, B, and C; we say that the variables A and B are "summed out" by q.

The path $q_N: N \to \underline{N}$ is in fact a multiple coend over all those variables in N which are summed out by q_N:

Lemma 2.5. *Let M be a functor into V and let $f: N \to M$ be a natural transformation which is, in particular, natural in all the repeated variables in N which are summed out by $q_N: N \to \underline{N}$. Then f factors as $g.q_N$ for a unique natural transformation $g: \underline{N} \to M$.*

Proof. This is by induction on the number of occurrences of $\underline{\otimes}$ in $\underline{N}$. If $\underline{N}$ contains no occurrence of $\underline{\otimes}$ the result is trivial; otherwise $\underline{N} = \underline{N}'\underline{\otimes}\underline{N}''$ and we can factor f in three steps:

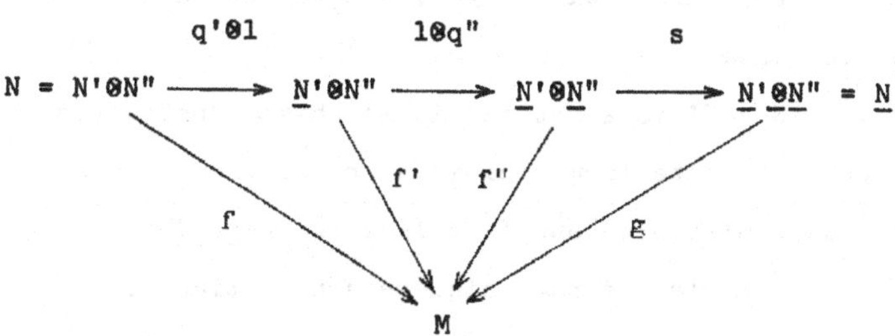

First consider the transform $\pi(f): N' \to [N'',M]$ of f under the tensor-hom adjunction isomorphism $\pi = Vp$ of V. By the induction hypothesis and routine naturality considerations, the diagram

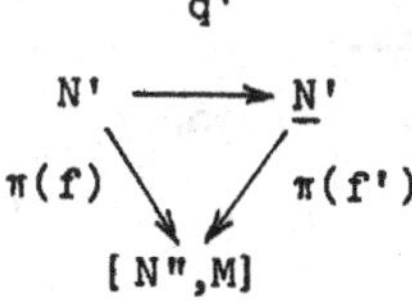

commutes for a unique morphism $\pi(f'): \underline{N}' \to [N'',M]$ where

$f': \underline{N}'\otimes N'' \to M$ is natural in all the variables not summed out by q'. Similarly f' factors as $f''.(1\otimes q'')$ for a unique morphism $f'': \underline{N}'\otimes\underline{N}'' \to M$ which is natural in all the variables not summed out by either of q' or q''. Finally, because s is a coend, f'' factors as g.s for a unique g: $\underline{N} \to M$ which is natural in all the remaining variables in $\underline{N}$ and M by Lemmas 2.2, 2.3, and 2.4. Combining these steps, we have that f factors through $q_N = s(q'\otimes q'')$ in the required manner.

When the transformation f in Lemma 2.5 is of the form $q'.n$ for a path $q': N' \to \underline{N}'$, the induced transformation g: $\underline{N} \to \underline{N}'$ is denoted by $\underline{n}$. Such induced transformations are a necessary part of the concept of a premonoidal category and we consider three relevant special cases below.

First, if n: N → N' is a natural isomorphism constructed entirely from the coherent data isomorphisms a, r, ℓ, c of V then $\underline{n}$: $\underline{N} \to \underline{N}'$ is a natural isomorphism and is called an <u>induced coherence isomorphism</u>. In view of the uniqueness assertion of Lemma 2.5, and the original coherence of a, r, ℓ, c, it is clear that induced coherence isomorphisms are coherent. In other words, the induced coherence isomorphism $\underline{n}$: $\underline{N} \to \underline{N}'$ is completely determined by the arrangement of $\underline{\otimes}$ in $\underline{N}$ and $\underline{N}'$; consequently we shall not label such isomorphisms.

Secondly, when $n = h\otimes k$: $S(A-)\otimes R(A-) \to S'(A-)\otimes R'(A-)$ for natural transformations h: S → S' and k: R → R', let us write

$\underline{h}\underline{\otimes}k$ for $\underline{h\otimes k}$. This not only makes the symbol $\underline{\otimes}$ $\mathcal{E}ns$-functorial whenever it is defined on objects, but also makes the coend s^A: $S(A-)\otimes R(A-) \to S(A-)\underline{\otimes}R(A-)$ $\mathcal{E}ns$-natural in S and R. Under reasonable conditions the same observations can be made at the $\mathcal{V}$-level.

If we restrict our attention to functors into $\mathcal{V}$ with <u>small domains</u> then the functors themselves may be regarded as extra variables. For example, let T: $\mathcal{A}^{op}\otimes\mathcal{A}\otimes\mathcal{B} \to \mathcal{V}$ be a functor with $\mathcal{A}$ and $\mathcal{B}$ small. Then $\int^A T(AAB)$ is canonically functorial in T and B for we can write T(AAB) = F(AATB) where F is the composite

$$\mathcal{A}^{op}\otimes\mathcal{A}\otimes([\mathcal{A}^{op}\otimes\mathcal{A}\otimes\mathcal{B},\mathcal{V}]\otimes\mathcal{B}) \cong [\mathcal{A}^{op}\otimes\mathcal{A}\otimes\mathcal{B},\mathcal{V}]\otimes(\mathcal{A}^{op}\otimes\mathcal{A}\otimes\mathcal{B}) \xrightarrow[E]{} \mathcal{V},$$

and where E is the evaluation functor defined in [1] §4. Similarly, if S(A-) and R(A-) are functors into $\mathcal{V}$ with small domains (and different variances in A) then $S(A-)\underline{\otimes}R(A-)$ is functorial in S and R in a unique way that makes s^A: $S(A-)\otimes R(A-) \to S(A-)\underline{\otimes}R(A-)$ natural in S and R.

Lastly, let S(A-) be a functor into $\mathcal{V}$ which is covariant in $A \in \mathcal{A}$. As part of the data for S, we have a family of morphisms S_{AB}: $\mathcal{A}(AB) \to [S(A-),S(B-)]$ which is natural in A and B and also in the extra variables in S. Transforming this family by the tensor-hom adjunction of $\mathcal{V}$, we get a transformation $\pi^{-1}(S_{AB})$: $\mathcal{A}(AB)\otimes S(A-) \to S(B-)$ which is natural in A and B and the extra variables in S. As a result of the generalised "higher" representation theorem (see [1], §3 and §5), this induces the <u>Yoneda isomorphism</u>

$$y_{S,B}\colon A(AB)\underline{\otimes}S(A-) \to S(B-).$$

By Lemma 2.2, we then have

Lemma 2.6. The Yoneda isomorphism $y_{S,A}$ is natural in A and in the extra variables in S; if the domain of S is small then it is natural in S.

The following diagram lemmas for the Yoneda isomorphism y are all proved using [3] Proposition II.7.4 which we shall refer to as the representation theorem. These lemmas are presented here in their most convenient forms for application in sections 3 and 4.

Lemma 2.7. Given functors $S\colon A \to V$ and $T\colon A^{op} \to V$ for which $SA\underline{\otimes}TA$ exists, the following diagram commutes:

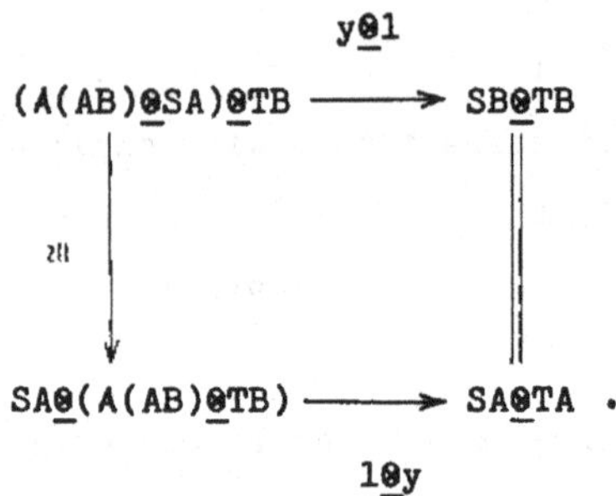

Proof. Replacing $\underline{\otimes}$ by $\otimes$, y by its definition, etc., we obtain a new diagram:

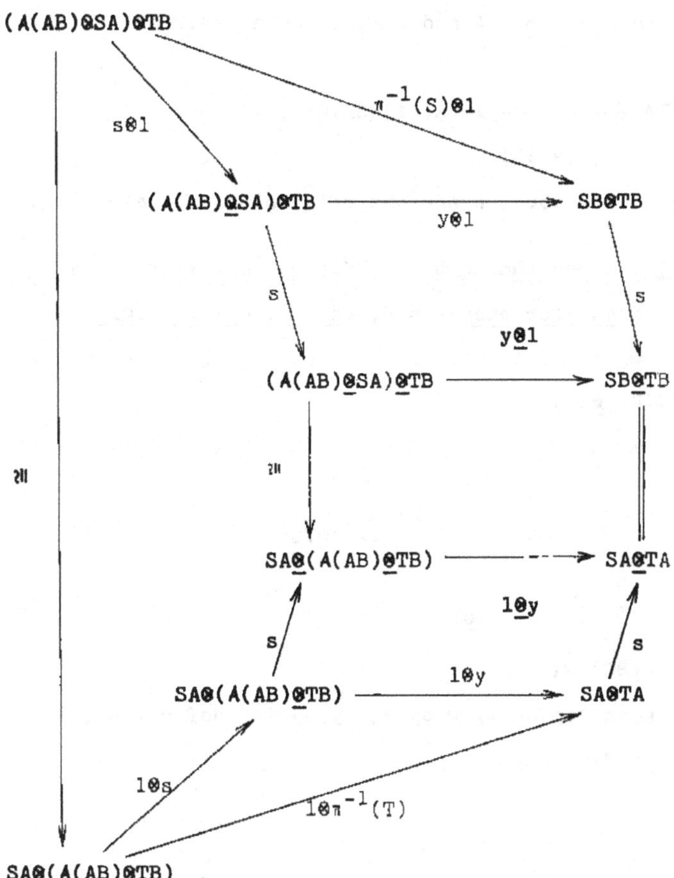

By Lemma 2.5, s(s⊗1) is a coend over A and B hence it suffices to prove that the exterior of this new diagram commutes for all A, B ∈ A. This is easily seen to be so on applying the representation theorem; put B = A and compose both exterior legs with

$$(I \otimes SA) \otimes TA \xrightarrow[(j_A \otimes 1) \otimes 1]{} (A(AA) \otimes SA) \otimes TA;$$

the resulting diagram commutes, hence the original one does.

<u>Lemma 2.8</u>. *Given functors* S: $A^{op} \otimes B \to V$ *and* T: $B^{op} \to V$ *for which* $S(AC)\underline{\otimes}TC$ *exists for each* $A \in A$, *the following diagram commutes for each* $A \in A$:

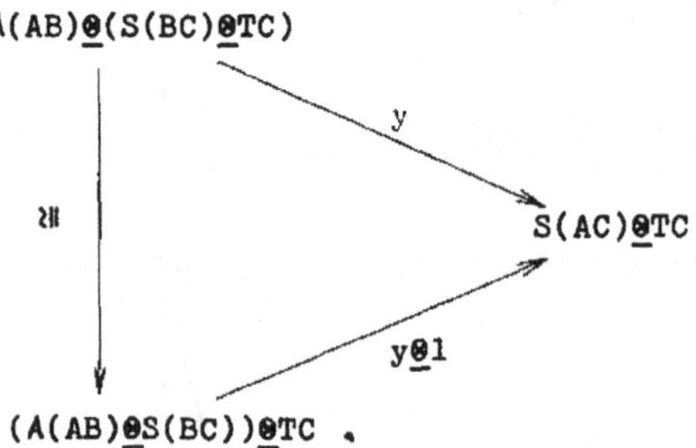

<u>Proof</u>. Again replacing $\underline{\otimes}$ by $\otimes$, y by its definition, etc., we obtain a new diagram:

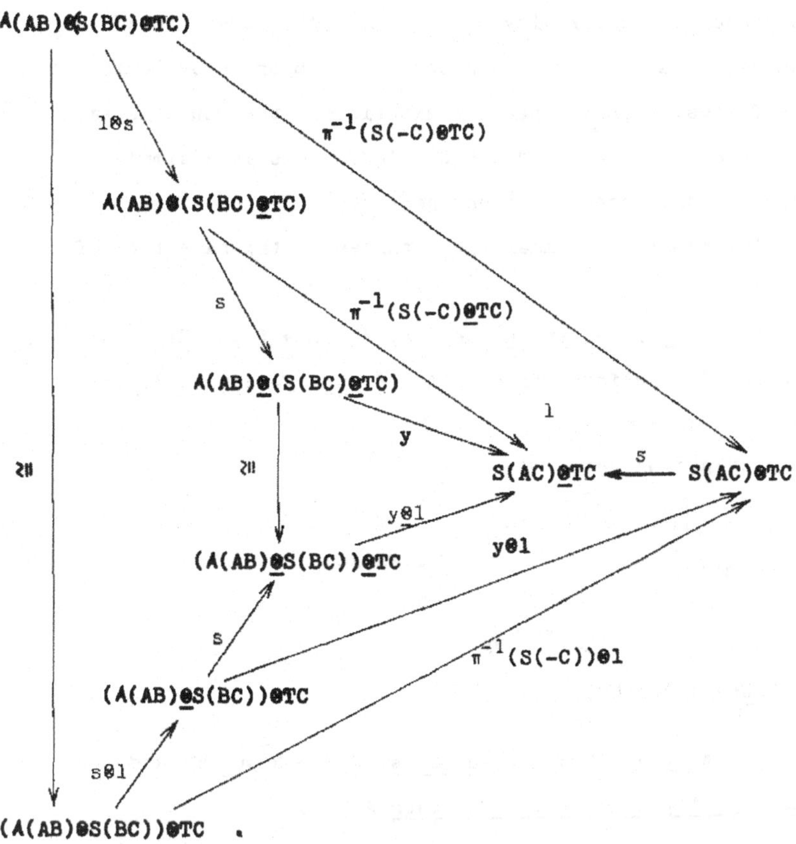

In this diagram the region labelled 1 commutes because it is the transform of the diagram

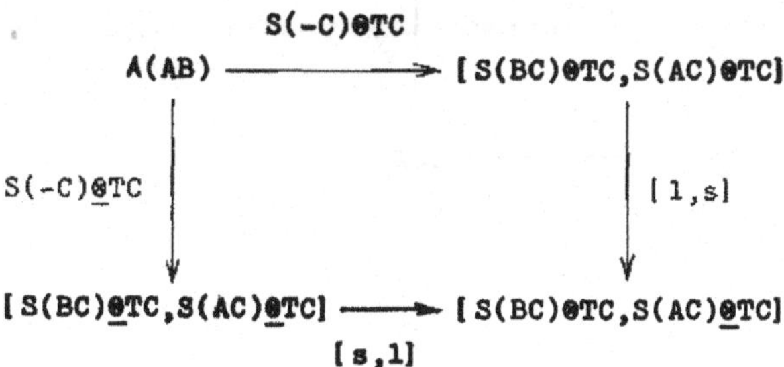

which expresses the naturality of $s = s^C$: $S(AC)\otimes TC \to S(AC)\underline{\otimes}TC$ in A. Hence, because $s(1\otimes s)$ is a coend over B and C by Lemma 2.5, it suffices to prove that the exterior of the new diagram commutes for all A, B $\in \mathcal{A}$ and C $\in \mathcal{B}$. Again this is a simple consequence of the representation theorem.

The remaining lemmas are obtained by the same type of argument.

<u>Lemma 2.9</u>. <u>Given functors</u> S: $\mathcal{A}^{op}\otimes\mathcal{B} \to \mathcal{V}$ <u>and</u> T: $\mathcal{B}^{op} \to \mathcal{V}$ <u>for which</u> $TC\underline{\otimes}S(AC)$ <u>exists for each</u> A $\in \mathcal{A}$, <u>the following diagram commutes for each</u> A $\in \mathcal{A}$:

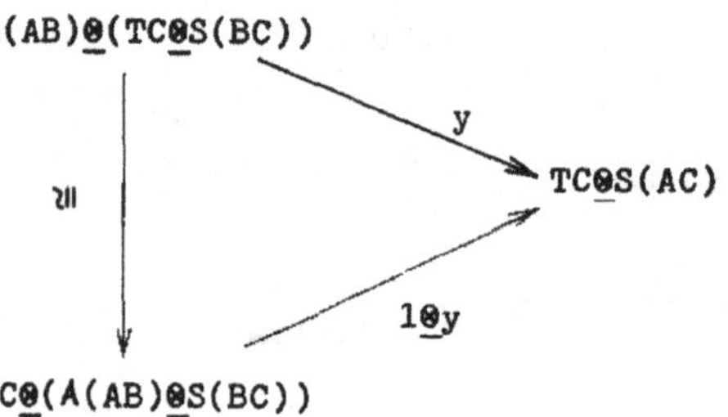

<u>Lemma 2.10</u>. <u>For any functors</u> S: $\mathcal{A} \to \mathcal{B}$ <u>and</u> T: $\mathcal{B}^{op} \to \mathcal{V}$ <u>the following diagram commutes for each</u> A $\in \mathcal{A}$:

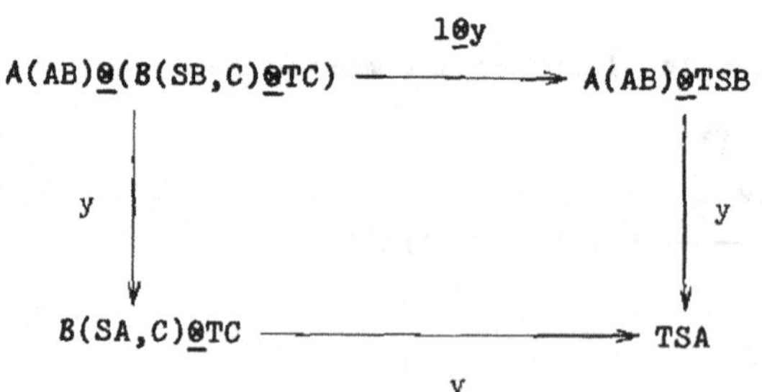

Lemma 2.11 *For any functor* T: $A \otimes B \to V$ *the following diagram commutes for all* B, D $\in A$:

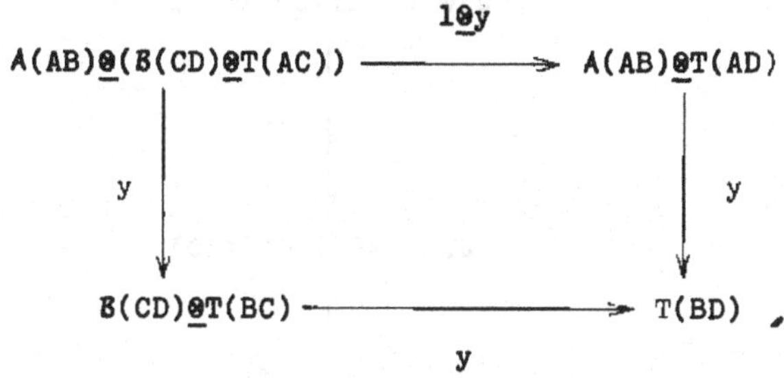

3. Premonoidal Categories

We emphasise again that, unless otherwise indicated, all concepts are relative to the given normalised symmetric monoidal closed category V.

Definition 3.1 A premonoidal category $P = (A,P,J,\lambda,\rho,\alpha)$ over V consists of

a category A,

a functor P: $A^{op} \otimes A^{op} \otimes A \to V$,

a functor J: $A \to V$,

and natural isomorphisms

$\lambda = \lambda_{AB}$: $JX \underline{\otimes} P(XAB) \to A(AB)$,

$\rho = \rho_{AB}$: $JX \underline{\otimes} P(AXB) \to A(AB)$,

$\alpha = \alpha_{ABCD}$: $P(ABX) \underline{\otimes} P(XCD) \to P(BCX) \underline{\otimes} P(AXD)$,

satisfying the following two axioms:

PC1. For all A,B,C ∈ A, the following diagram commutes:

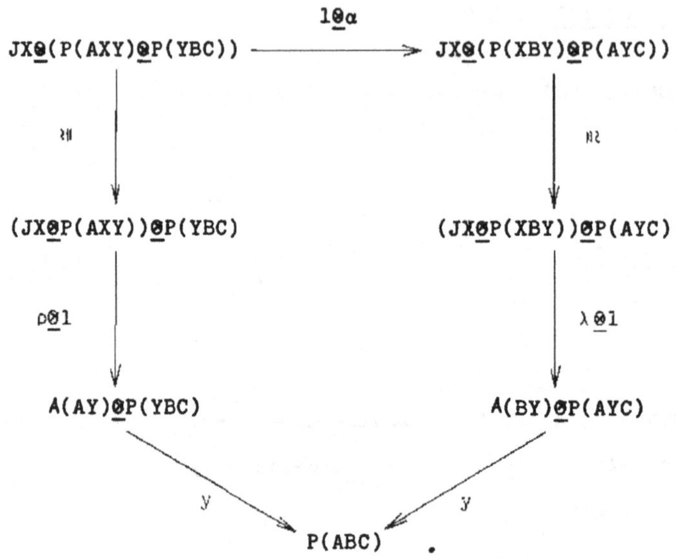

PC2. For all A,B,C,D,E ∈ A, the following diagram commutes:

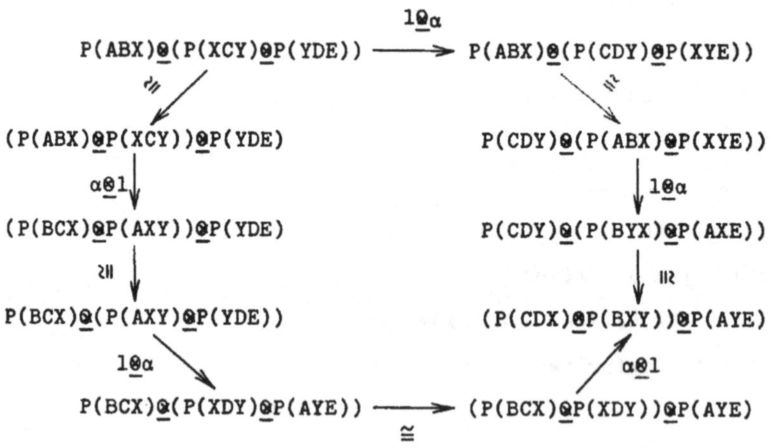

Remark 3.2 It is assumed in the definition that the requisite $\underline{\otimes}$'s exist for the given A, P, and J. They do so, by hypothesis on V, when A is small. They also exist whenever $P(AB-): A \to V$ and $J: A \to V$ are representable for all $A,B \in A$.

In the remainder of this section we will suppose that A is small and show that each premonoidal structure P on A "extends" to a biclosed structure $[P,V]$ on the functor category $[A,V]$. For the monoidal part define a tensor-product $*: [A,V]\otimes[A,V] \to [A,V]$ by

(3.1) $$S*T = \int^A SA\otimes\int^B TB\otimes P(AB-) = SA\underline{\otimes}(TB\underline{\otimes}P(AB-))$$

for all $S,T \in [A,V]$; this expression is canonically functorial in S and T by the considerations of section 2. Next, let $J \in [A,V]$ be the identity-object of $*$, and define natural isomorphisms $\ell^* = \ell^*_T: J*T \to T$ and $r^* = r^*_T: T*J \to J$ as the respective composites

$$J*T = JX\underline{\otimes}(TA\underline{\otimes}P(XA-)) \cong (JX\underline{\otimes}P(XA-))\underline{\otimes}TA \xrightarrow[\lambda\underline{\otimes}1]{} A(A-)\underline{\otimes}TA \xrightarrow[y]{} T$$

and

$$T*J = TA\underline{\otimes}(JX\underline{\otimes}P(AX-)) \cong (JX\underline{\otimes}P(AX-))\underline{\otimes}TA \xrightarrow[\rho\otimes1]{} A(A-)\underline{\otimes}TA \xrightarrow[y]{} T.$$

Lastly, define a natural isomorphism $a^* = a^*_{RST}: (R*S)*T \to R*(S*T)$ as the composite

$$
\begin{aligned}
(R*S)*T &= (RA\underline{\otimes}(SB\underline{\otimes}P(ABX)))\underline{\otimes}(TC\underline{\otimes}P(XC-)) \\
&\cong RA\underline{\otimes}(SB\underline{\otimes}(TC\underline{\otimes}(P(ABX)\underline{\otimes}P(XC-)))) \\
&\xrightarrow[1\underline{\otimes}(1\underline{\otimes}(1\underline{\otimes}\alpha))]{} RA\underline{\otimes}(SB\underline{\otimes}(TC\underline{\otimes}(P(BCX)\underline{\otimes}P(AX-)))) \\
&\cong RA\underline{\otimes}((SB\underline{\otimes}(TC\underline{\otimes}P(BCX)))\underline{\otimes}P(AX-)) \\
&= R*(S*T).
\end{aligned}
$$

Then ℓ^*, r^*, and a^* are natural by Lemmas 2.5 and 2.6.

<u>Theorem 3.3</u> $[P,V] = ([A,V],*,J,\ell^*,r^*,a^*)$ <u>is a monoidal category admitting a biclosed structure.</u>

<u>Proof</u> First, to show that $[P,V]$ is a monoidal category, we need to prove PC1 $\Rightarrow$ MC2 and PC2 $\Rightarrow$ MC3. The first of these is obtained by considering diagram (3.2) in which the exterior commutes by PC1; 1, 2, and 3 commute by the definitions of $*,r^*,a^*$, and ℓ^*; 4, 5, 6, 7, 8, and 9 commute by the naturality and coherence of the induced coherence isomorphisms (Lemma 2.5 and the succeeding remarks); 10 and 11 commute by Lemma 2.7; and 12 commutes by Lemma 2.9. The proof of PC2 $\Rightarrow$ MC3 requires a diagram that is too large for the space available but, apart from the definitions of $*$ and a^*, uses only the naturality and coherence of the induced coherence isomorphisms involved.

To complete the structure on $[A,V]$ to that of a biclosed category, consider the composite isomorphism:

Diagram (3.2)

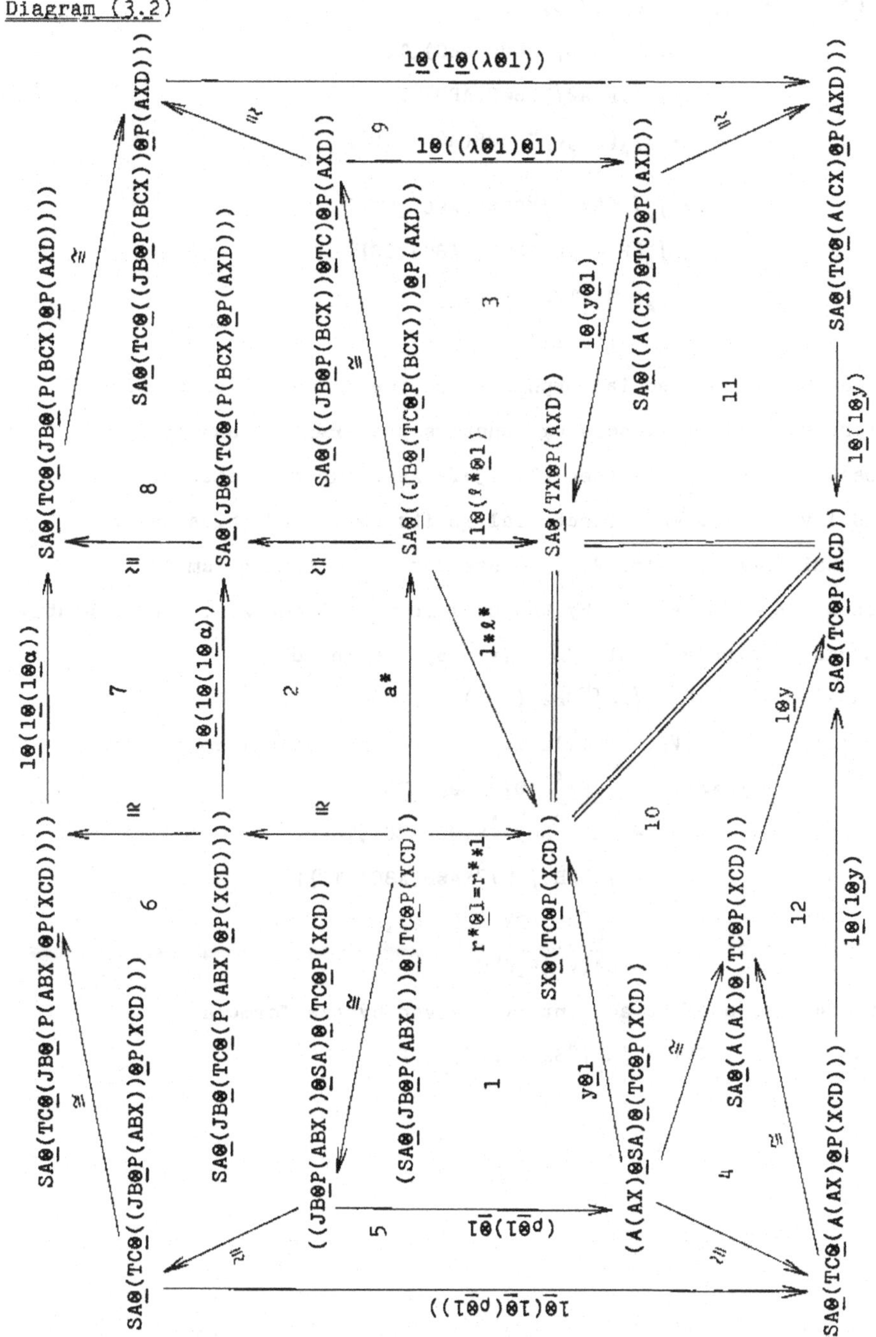

$$
\begin{aligned}
[A,V](R*S,T) &= \int_C[(R*S)C,TC] \\
&= \int_C[\int^A RA\otimes\int^B SB\otimes P(ABC),TC] \\
&\cong \int_C\int_A[RA\otimes\int^B SB\otimes P(ABC),TC] \\
&\xrightarrow[\int\int p]{} \int_C\int_A[RA,[\int^B SB\otimes P(ABC),TC]] \\
&\cong \int_A\int_C[RA,[\int^B SB\otimes P(ABC),TC]] \\
&\cong \int_A[RA,\int_C[\int^B SB\otimes P(ABC),TC]] \\
&= \int_A[RA,(T/S)A] \text{ say,} \\
&= [A,V](R,T/S),
\end{aligned}
$$

where the unlabelled isomorphisms are the canonical ones which assert that limit-preserving functors preserve ends and that repeated ends commute (see [1] §3). Assuming that each of the ends involved is made functorial in its extra variables using the dual form of Lemma 2.1, we see that each isomorphism is natural in R, S, and T, by the dual form of Lemma 2.2. Consequently $-*S$ has a right adjoint $-/S$, given by the formula

$$(3.3) \qquad T/S = \int_C[\int^B SB\otimes P(-BC),TC]$$

for all $S,T \in [A,V]$. Similarly we have the natural composite

$$
\begin{aligned}
[A,V](S*R,T) &= \int_C[\int^A SA\otimes\int^B RB\otimes P(ABC),TC] \\
&\cong \int_C[\int^B RB\otimes\int^A SA\otimes P(ABC),TC] \\
&\cong \int_B[RB,\int_C[\int^A SA\otimes P(ABC),TC]] \\
&= \int_B[RB,(S\backslash T)B] \text{ say,} \\
&= [A,V](R,S\backslash T).
\end{aligned}
$$

Thus $S*-$ has a right adjoint $S\backslash -$, given by the formula

$$(3.4) \qquad S\backslash T = \int_C[\int^A SA\otimes P(A-C),TC]$$

for all $S,T \in [A,V]$. This completes the proof.

Definition 3.4 A symmetry for the premonoidal category P is a natural isomorphism

$$\sigma = \sigma_{ABC}: P(ABC) \to P(BAC)$$

satisfying the following two axioms:

PC3. $\sigma^2 = 1$

PC4. For all $A,B,C,D \in A$, the following diagram commutes:

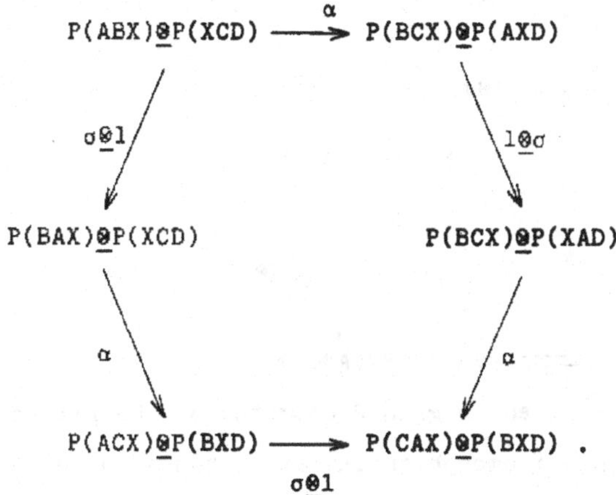

Remark 3.5 This definition does not, of course, require A to be small.

It remains to be shown that $[P,V]$ admits a symmetric monoidal closed structure whenever P has a symmetry. For this, define a natural isomorphism $c^* = c^*_{ST}: S*T \to T*S$ as the composite

$$S*T = SA\underline{\otimes}(TB\underline{\otimes}P(AB-)) \cong TB\underline{\otimes}(SA\underline{\otimes}P(AB-))$$

$$\xrightarrow[1\underline{\otimes}(1\underline{\otimes}\sigma)]{} TB\underline{\otimes}(SA\underline{\otimes}P(BA-)) = T*S.$$

Again, the naturality of c* is a consequence of Lemma 2.5.

<u>Theorem 3.6</u> <u>If</u> σ <u>is a symmetry for</u> P <u>then</u> c* <u>is a symmetry for</u> $[P,V]$.

<u>Proof</u> To prove PC3 ⇒ MC6 consider diagram (3.5):

(3.5)

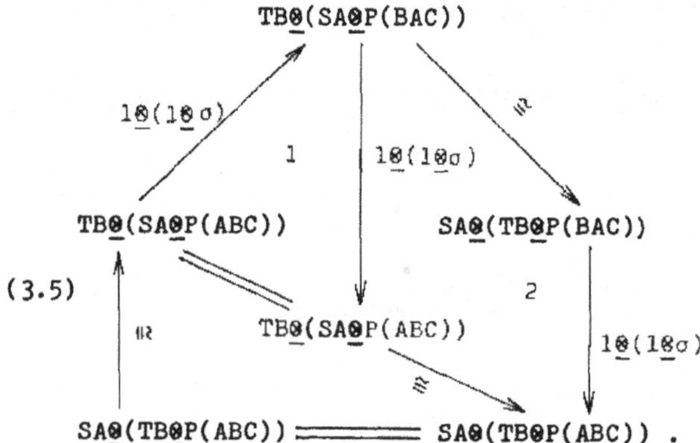

Region 1 commutes by PC3, and region 2 commutes by the naturality of the induced coherence isomorphism involved; hence the exterior commutes and so, by definition of c*, MC6 is satisfied. To prove PC4 ⇒ MC7 consider diagram (3.6), in which the exterior commutes by PC4; 1, 2, and 3 commute by the definitions of * and c*; 4, 5, and 6 commute by the definition of a*; and 7, 8, 9, and 10 commute by the naturality and coherence of induced coherence isomorphisms.

Diagram (3.6)

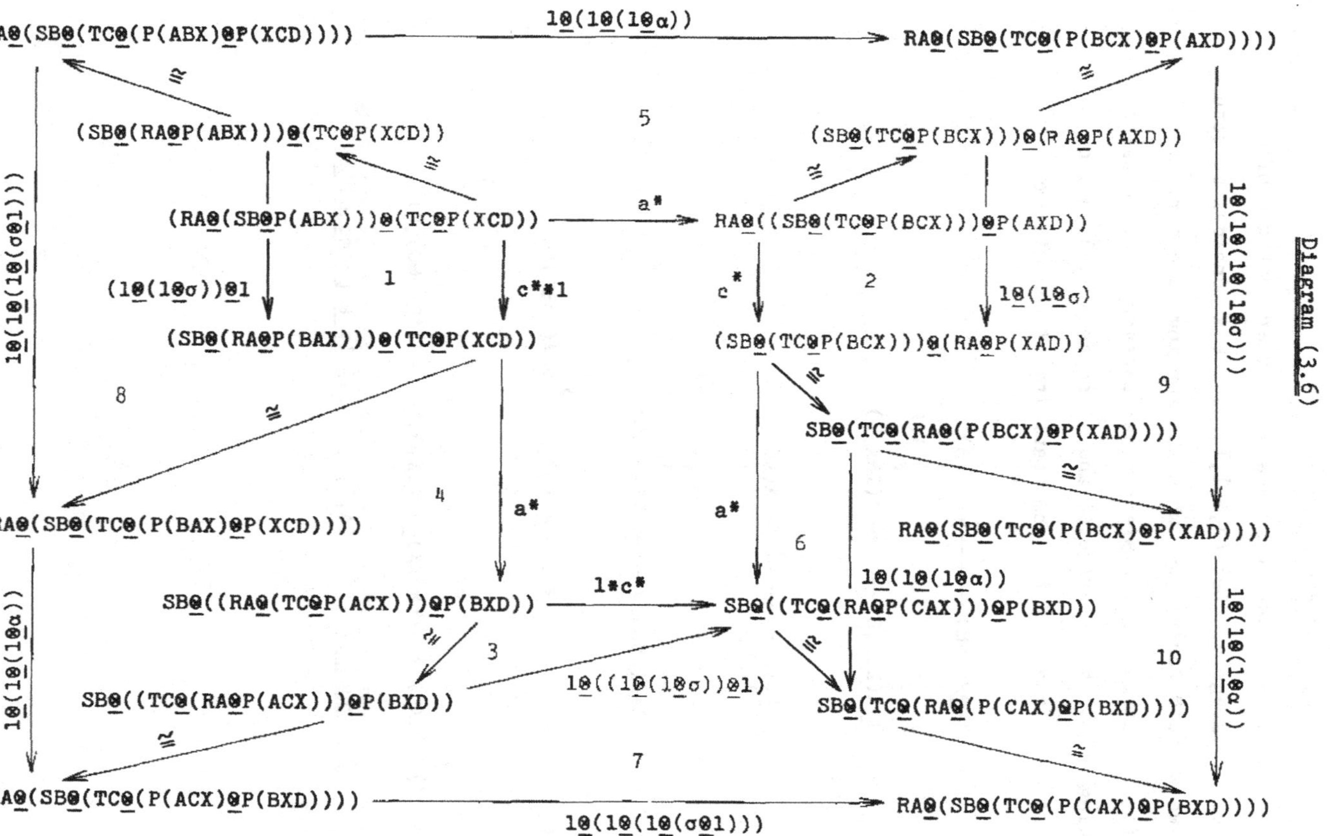
RA⊗(SB⊗(TC⊗(P(ABX)⊗P(XCD))))
1⊗(1⊗(1⊗α))
RA⊗(SB⊗(TC⊗(P(BCX)⊗P(AXD))))
(SB⊗(RA⊗P(ABX)))⊗(TC⊗P(XCD))
5
(SB⊗(TC⊗P(BCX)))⊗(RA⊗P(AXD))
(RA⊗(SB⊗P(ABX)))⊗(TC⊗P(XCD))
a*
RA⊗((SB⊗(TC⊗P(BCX)))⊗P(AXD))
1⊗(1⊗(1⊗(σ⊗1)))
(1⊗(1⊗σ))⊗1
1
c*⊗1
c*
2
1⊗(1⊗σ)
1⊗(1⊗(1⊗(1⊗σ)))
(SB⊗(RA⊗P(BAX)))⊗(TC⊗P(XCD))
(SB⊗(TC⊗P(BCX)))⊗(RA⊗P(XAD))
8
9
SB⊗(TC⊗(RA⊗(P(BCX)⊗P(XAD))))
4
a*
a*
RA⊗(SB⊗(TC⊗(P(BAX)⊗P(XCD))))
6
RA⊗(SB⊗(TC⊗(P(BCX)⊗P(XAD))))
1⊗(1⊗(1⊗α))
SB⊗((RA⊗(TC⊗P(ACX)))⊗P(BXD))
1*c*
SB⊗((TC⊗(RA⊗P(CAX)))⊗P(BXD))
1⊗(1⊗α)
3
10
1⊗(1⊗(1⊗α))
1⊗((1⊗(1⊗σ))⊗1)
SB⊗((TC⊗(RA⊗P(ACX)))⊗P(BXD))
SB⊗(TC⊗(RA⊗(P(CAX)⊗P(BXD))))
7
RA⊗(SB⊗(TC⊗(P(ACX)⊗P(BXD))))
RA⊗(SB⊗(TC⊗(P(CAX)⊗P(BXD))))
1⊗(1⊗(1⊗(σ⊗1)))

4. Monoidal Categories

A monoidal category is a particular instance of a premonoidal category. Let $(\mathcal{A},\bar{\otimes},\bar{I},\bar{\ell},\bar{r},\bar{a})$ be a monoidal structure on $\mathcal{A}$; we write $\otimes$ for $\bar{\otimes}$, etc. when the meaning is clear. The data for the corresponding premonoidal category $\mathcal{P}$ are obtained by taking $P(ABC)$ to be $\mathcal{A}(A\otimes B,C)$ and JA to be $\mathcal{A}(I,A)$, and by defining λ, ρ, and α by the commutativity of the diagrams

$$\begin{array}{ccc}
JX\underline{\otimes}P(XAB) & \xrightarrow{\;\lambda\;} & \mathcal{A}(AB) \\
\Vert & & \big\downarrow{\scriptstyle \mathcal{A}(\ell,1)} \\
\mathcal{A}(IX)\underline{\otimes}\mathcal{A}(X\otimes A,B) & \xrightarrow[\;y\;]{} & \mathcal{A}(I\otimes A,B)
\end{array} \quad ,$$

$$\begin{array}{ccc}
JX\underline{\otimes}P(AXB) & \xrightarrow{\;\rho\;} & \mathcal{A}(AB) \\
\Vert & & \big\downarrow{\scriptstyle \mathcal{A}(r,1)} \\
\mathcal{A}(IX)\underline{\otimes}\mathcal{A}(A\otimes X,B) & \xrightarrow[\;y\;]{} & \mathcal{A}(A\otimes I,B)
\end{array} \quad ,$$

and

$$\begin{array}{ccccc}
P(ABX)\underline{\otimes}P(XCD) & = & \mathcal{A}(A\otimes B,X)\underline{\otimes}\mathcal{A}(X\otimes C,D) & \xrightarrow{\;y\;} & \mathcal{A}((A\otimes B)\otimes C,D) \\
{\scriptstyle \alpha}\big\downarrow & & & & \big\uparrow{\scriptstyle \mathcal{A}(a,1)} \\
P(BCX)\underline{\otimes}P(AXD) & = & \mathcal{A}(B\otimes C,X)\underline{\otimes}\mathcal{A}(A\otimes X,D) & \xrightarrow[\;y\;]{} & \mathcal{A}(A\otimes(B\otimes C),D)
\end{array} \quad .$$

Theorem 4.1 $\mathcal{P} = (\mathcal{A},P,J,\lambda,\rho,\alpha)$ *is a premonoidal category.*

Diagram (4.1)

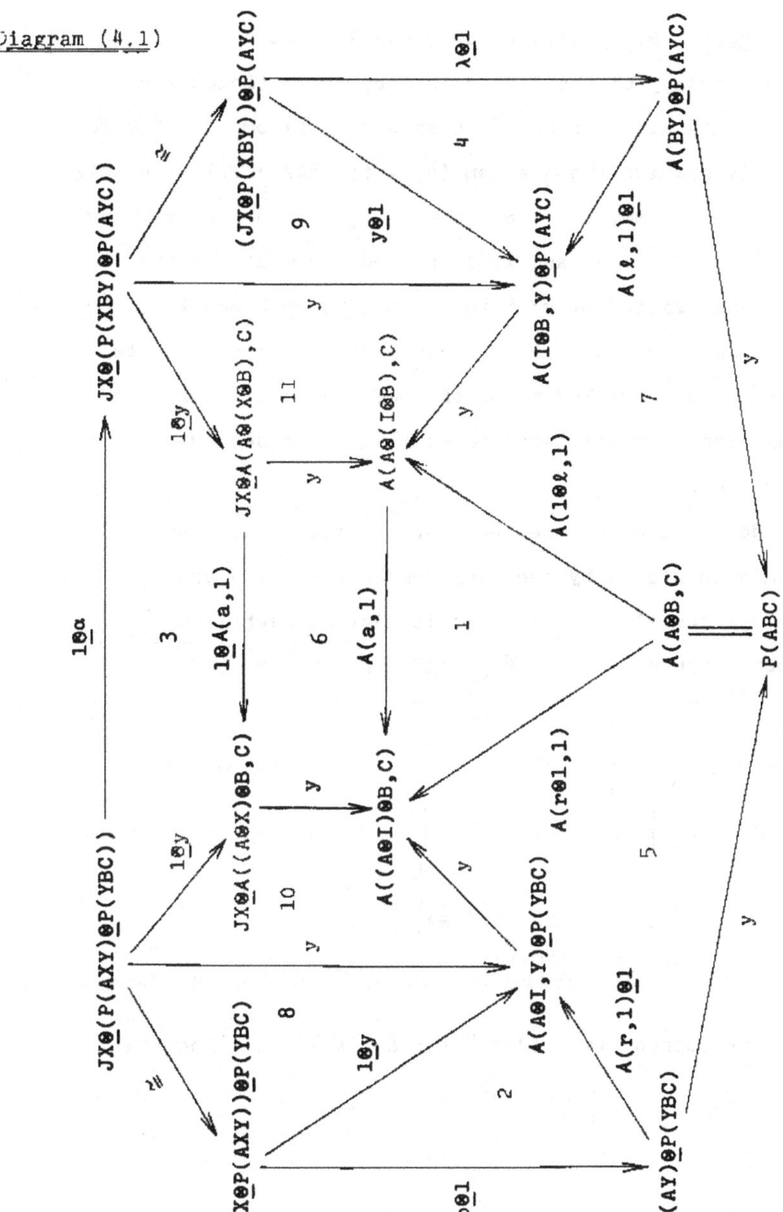
JX⊗(P(AXY)⊗P(YBC))
1⊗α
JX⊗(P(XBY)⊗P(AYC))
≅
(JX⊗P(XBY))⊗P(AYC)
λ⊗1
A(BY)⊗P(AYC)
4
9
y⊗1
A(I⊗B,Y)⊗P(AYC)
A(ℓ,1)⊗1
y
1⊗y
JX⊗A(A⊗(X⊗B),C)
11
A(A⊗(I⊗B),C)
7
A(1⊗ℓ,1)
3
1⊗A(a,1)
6
A(a,1)
1
A(A⊗B,C)
P(ABC)
JX⊗A((A⊗X)⊗B,C)
A((A⊗I)⊗B,C)
A(r⊗1,1)
5
10
8
A(A⊗I,Y)⊗P(YBC)
(JX⊗P(AXY))⊗P(YBC)
A(r,1)⊗1
2
ρ⊗1
A(AY)⊗P(YBC)

<u>Proof</u> Naturality of the data λ, ρ, and α follows from the naturality of the Yoneda isomorphism y (Lemma 2.6) and of $\bar{\ell}$, $\bar{r}$, and $\bar{a}$. It remains to establish axioms PC1 and PC2; we shall only provide the diagram (4.1) for MC2 $\Rightarrow$ PC1. In this diagram region 1 commutes by axiom MC2; 2, 3, and 4 commute by the definitions of ρ, α, and λ; 5, 6, and 7 commute by the evident $\mathcal{E}ns$-naturality of y; 8 and 9 commute by Lemma 2.8; and 10 and 11 commute by Lemma 2.10. Hence the exterior commutes, as required. To prove MC3 $\Rightarrow$ PC2 one requires a similar (but larger) diagram; the only additional results needed are Lemmas 2.9 and 2.11.

Monoidal structures on A are in fact characterised among premonoidal ones by the representability of J and of $P(AB-): A \to V$ for all $A,B \in A$. It is also straightforward to verify that a symmetry $\bar{c}$ for $\bar{\otimes}$ provides a symmetry σ for P, defined by:

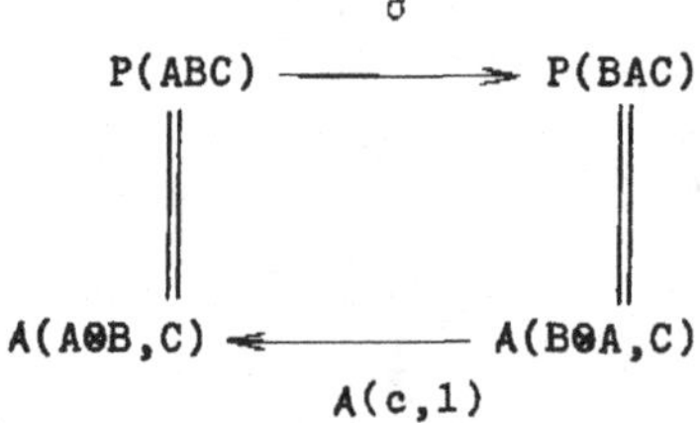

commutes.

Now suppose that $(A,\bar{\otimes},\bar{I},\bar{\ell},\bar{r},\bar{a})$ is a small monoidal category.

Applying the generalised representation theorem to the internal-hom formulas (3.3) and (3.4) for $[\mathcal{P},\mathcal{V}]$, we get:

$$\begin{aligned}(T/S)A &= \int_C[\int^B SB\otimes P(ABC),TC] \\ &= \int_C[\int^B SB\otimes A(A\otimes B,C),TC] \\ &\cong \int_B[SB, \int_C[A(A\otimes B,C),TC]] \\ &\cong \int_B[SB,T(A\otimes B)],\end{aligned}$$

and

$$\begin{aligned}(S\backslash T)A &= \int_C[\int^B SB\otimes P(BAC),TC] \\ &\cong \int_B[SB,T(B\otimes A)],\end{aligned}$$

for all $S,T \in [\mathcal{A},\mathcal{V}]$ and $A \in \mathcal{A}$. If, in addition, the functor $P(A-C)\colon \mathcal{A}^{op} \to \mathcal{V}$ admits a representation $\mathcal{A}(-,A\backslash C)\colon \mathcal{A}^{op} \to \mathcal{V}$ for all $A,C \in \mathcal{A}$ then, on applying the generalised representation theorem to the tensor-product formula (3.1) for $[\mathcal{P},\mathcal{V}]$, we get a convolution formula:

$$\begin{aligned}(S*T)C &= \int^A SA\otimes\int^B TB\otimes P(ABC) \\ &= \int^A SA\otimes\int^B TB\otimes A(B,A\backslash C) \\ &\cong \int^A SA\otimes\int^B A(B,A\backslash C)\otimes TB \\ &\cong \int^A SA\otimes T(A\backslash C)\end{aligned}$$

for all $S,T \in [\mathcal{A},\mathcal{V}]$ and $C \in \mathcal{A}$.

Here are three examples of closed functor categories that arise in this way. For certain choices of the ground category $\mathcal{V}$ (e.g. $\mathcal{V} = \mathcal{E}ns$ and $\mathcal{V} = \mathcal{A}b$) these examples are quite well known.

<u>Example 4.2</u> If A is a category with only one object $\overline{I}$ whose endomorphism - monoid (i.e. endomorphism - algebra) $\{M = A(\overline{I},\overline{I})$, $\mu: M \otimes M \to M$, $\eta: I \to M\}$ is <u>commutative</u> then we can, by [3] Proposition III.4.2, define a functor $\overline{\otimes}: A \otimes A \to A$ with the data $\overline{I} \overline{\otimes} \overline{I} = \overline{I}$ and $\mu: M \otimes M \to M$. Taking each of $\overline{\ell}$, $\overline{r}$, $\overline{a}$, and $\overline{c}$ to be the identity transformation of the identity functor on A, we see that $(A, \overline{\otimes}, \overline{I}, \overline{\ell}, \overline{r}, \overline{a}, \overline{c})$ is a symmetric monoidal category. In this example we may also take $\overline{I} \backslash \overline{I} = \overline{I} = \overline{I} \overline{\otimes} \overline{I}$ because $A^{op} = A$; it is then easy to check that $[P,V]$ is the category of M-modules with the usual tensor-product and internal-hom.

<u>Example 4.3</u> Let $V = Ens$ and let A be a (finitary) <u>commutative theory</u> in the sense of Linton [9]. Recall that commutativity of A means that, for each m-ary operation $\mu \in A(m,1)$ and n-ary operation $\nu \in A(n,1)$, the following diagram commutes:

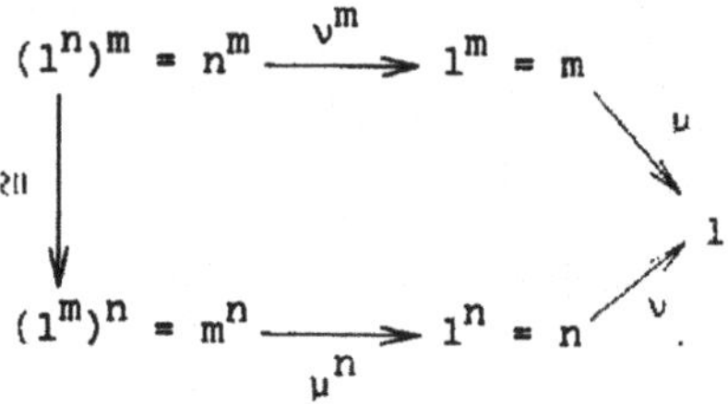

This condition is sufficient for the existence of a functor $\overline{\otimes}: A \times A \to A$ defined by $m \overline{\otimes} n = n^m$ and $\mu \overline{\otimes} \nu = \mu . \nu^m$. Let $\overline{I} = 1$, let $\overline{\ell}$, $\overline{r}$, and $\overline{a}$ be the appropriate identity isomorphisms, and let $\overline{c}$ be the canonical "switching" isomorphism shown in the diagram. These data provide A with the structure of a symmetric monoidal category which, in turn, yields a symmetric monoidal closed structure $[P,Ens]$ on the category $[A,Ens]$ of A-prealgebras. When this structure is restricted to the full subcategory of $[A,Ens]$ determined by the A-algebras, we obtain the usual symmetric monoidal closed category of algebras over the commutative theory A.

In fact the above assertions remain valid if we replace Ens by any cartesian closed V (having small limits and colimits); a finitary V-theory is a V-category A having for objects the non-negative integers 0, 1, ... , n, ... and having the property that $n = 1^n$ in A_0 and $A(m,n) \cong A(m,1)^n$ in V_0 for all $m,n \in A$; the definition of commutativity is the V-analogue of the above and can easily be deduced from [3] III Proposition 4.2. For still further generalisation see Kock [6] and [7].

Example 4.4 First let us note that the monoidal closed normalisation functor $V: V \to Ens$ has a monoidal closed left adjoint $F: Ens \to V$ which sends a set X to the copower $\sum_X I$ in V_0 of X copies of I (see [5] §5 for related generalities). The induced monoidal functor $F_{\#}: Ens_{\#} \to V_{\#}$ (of [3] Proposition III.3.6) sends the Ens-category C to the category $F_{\#}C$ whose objects are those of C and whose hom-objects are given by $(F_{\#}C)(AB) = \sum_{C(AB)} I$ in V.

Now let C be the discrete Ens-category whose object-set is the abelian group of integers $\mathbb{Z}$. Putting $A = F_{\#}C$, defining

$m\bar{\otimes}n = m + n$, $n\backslash m = m - n$, $\bar{I} = 0 \in \mathbb{Z}$, and taking $\bar{\ell}$, $\bar{r}$, $\bar{a}$, and $\bar{c}$ to be the appropriate identity transformations, we obtain a symmetric monoidal closed structure on A. Because C is a discrete $\mathcal{E}ns$-category, $\int^{\cdot}$ reduces to $\sum$ and $\int_{\cdot}$ to Π in V_0 so that the resulting symmetric monoidal closed structure on the category $[A,V]$ of $\mathbb{Z}$-graded objects in V is given by

$$(X*Y)_m = \int^n X_n \otimes Y_{n\backslash m} = \sum_{n\in\mathbb{Z}} X_n \otimes Y_{m-n}$$

and

$$(Y/X)_m = \int_n [X_n, Y_{m\bar{\otimes}n}] = \prod_{n\in\mathbb{Z}} [X_n, Y_{m+n}]$$

for all $X,Y \in [A,V]$ and $m \in \mathbb{Z}$.

5. Other Examples

Another type of premonoidal structure arises when a category A has the structure of a comonoid in the monoidal category $V_\#$. Such a comonoid consists of a comultiplication functor $\delta\colon A \to A\otimes A$ and a counit functor $\varepsilon\colon A \to I$ satisfying the following coassociative and left and right counit laws (in which the unlabelled isomorphisms are the data isomorphisms of $V_\#$):

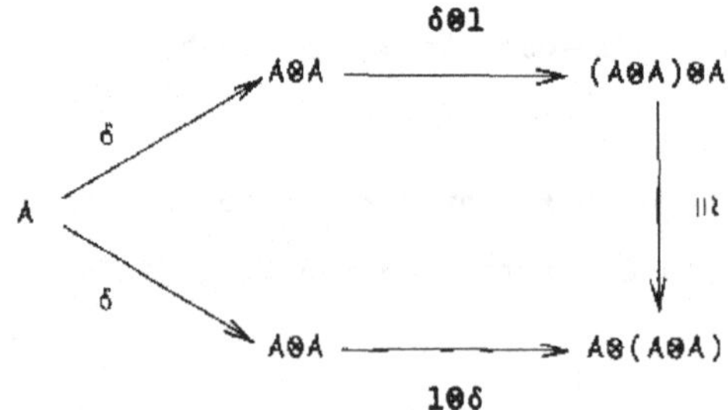

commutes, and

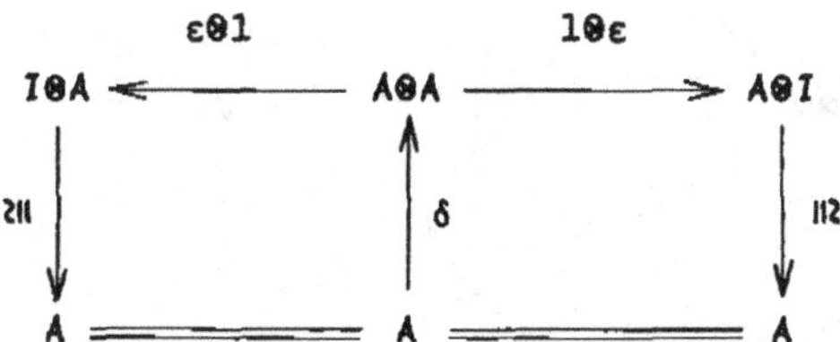

commute. These laws imply that δ sends an object $A \in \mathcal{A}$ to $(A,A) \in \mathcal{A}\otimes\mathcal{A}$ and that the morphisms δ_{AB}: $\mathcal{A}(AB) \to \mathcal{A}(AB)\otimes\mathcal{A}(AB)$ and ε_{AB}: $\mathcal{A}(AB) \to I$ provide a comonoid structure on the hom-object $\mathcal{A}(AB)$ for each pair of objects $A,B \in \mathcal{A}$.

A premonoidal structure P is now defined on $\mathcal{A}$ with the following data. Take P and J to be the respective composites

$$(\mathcal{A}^{op}\otimes\mathcal{A}^{op})\otimes\mathcal{A} \xrightarrow{1\otimes\delta} (\mathcal{A}^{op}\otimes\mathcal{A}^{op})\otimes(\mathcal{A}\otimes\mathcal{A}) \xrightarrow{\mathrm{Hom}(\mathcal{A}\otimes\mathcal{A})} \mathcal{V}$$

and

$$\mathcal{A} \xrightarrow{\varepsilon} \mathcal{I} \xrightarrow{I} \mathcal{V},$$

so that, on objects, $P(ABC) = \mathcal{A}(AC)\otimes\mathcal{A}(BC)$ and $JA = I$. Define $\lambda = \lambda_{AB}$ as the composite

$$JX\underline{\otimes}P(XAB) = JX\underline{\otimes}(\mathcal{A}(XB)\otimes\mathcal{A}(AB)) \cong (\mathcal{A}(XB)\underline{\otimes}JX)\otimes\mathcal{A}(AB)$$

$$\xrightarrow[y\otimes 1]{} JB\otimes\mathcal{A}(AB) \xrightarrow[\ell]{} \mathcal{A}(AB),$$

noting that, for each $A \in \mathcal{A}$, the last arrow is actually the B-component of the horizontal composite

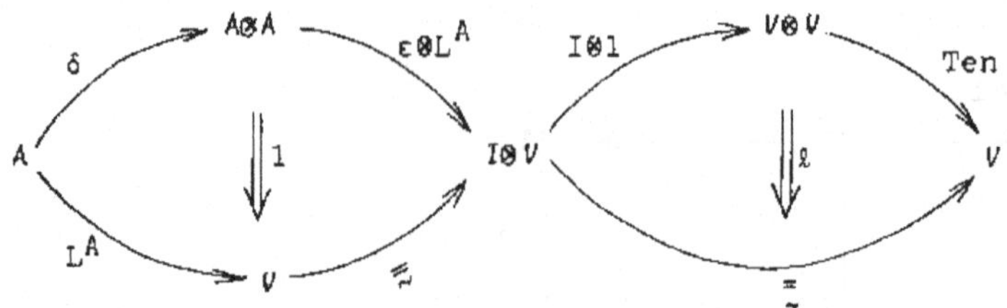

of natural transformations. Similarly, define $\rho = \rho_{AB}$ as the composite

$$JX\underline{\otimes}P(AXB) = JX\underline{\otimes}(A(AB)\otimes A(XB)) \cong A(AB)\otimes(A(XB)\underline{\otimes}JX)$$

$$\xrightarrow[1\otimes y]{} A(AB)\otimes JB \xrightarrow[r]{} A(AB),$$

and $\alpha = \alpha_{ABCD}$ as the composite

$$P(ABX)\underline{\otimes}P(XCD) = P(ABX)\underline{\otimes}(A(XD)\otimes A(CD)) \cong (A(XD)\underline{\otimes}P(ABX))\otimes A(CD)$$

$$\xrightarrow[y\otimes 1]{} P(ABD)\otimes A(CD) = (A(AD)\otimes A(BD))\otimes A(CD)$$

$$\xrightarrow[a]{} A(AD)\otimes(A(BD)\otimes A(CD)) = A(AD)\otimes P(BCD)$$

$$\xrightarrow[1\otimes y^{-1}]{} A(AD)\otimes(A(XD)\underline{\otimes}P(BCX)) \cong P(BCX)\underline{\otimes}(A(AD)\otimes A(XD))$$

$$= P(BCX)\underline{\otimes}P(AXD).$$

Furthermore, if the comultiplication δ is commutative we can define a symmetry $\sigma = \sigma_{ABC}$ for P as

$$P(ABC) = A(AC)\otimes A(BC) \xrightarrow[c]{} A(BC)\otimes A(AC) = P(BAC).$$

We now suppose that $\mathcal{A}$ is small and, as in section 4, use the generalised representation theorem to reduce the tensor-product and internal-hom formulas (3.1), (3.3), and (3.4) for $[\mathcal{P},\mathcal{V}]$:

$$\begin{aligned}(S*T)C &= \textstyle\int^{A} SA\otimes\int^{B} TB\otimes P(ABC)\\ &= \textstyle\int^{A} SA\otimes\int^{B} TB\otimes(\mathcal{A}(AC)\otimes\mathcal{A}(BC))\\ &\cong (\textstyle\int^{A} SA\otimes\mathcal{A}(AC))\otimes(\int^{B} TB\otimes\mathcal{A}(BC))\\ &\cong (\textstyle\int^{A} \mathcal{A}(AC)\otimes SA)\otimes(\int^{B} \mathcal{A}(BC)\otimes TB)\\ &\cong SC\otimes TC\end{aligned}$$

for all $S,T \in [\mathcal{A},\mathcal{V}]$ and $C \in \mathcal{A}$, and

$$\begin{aligned}(T/S)A &= \textstyle\int_{C}[\int^{B} SB\otimes P(ABC),TC]\\ &\cong \textstyle\int_{C}[\int^{B} P(ABC)\otimes SB,TC]\\ &= \textstyle\int_{C}[\int^{B} (\mathcal{A}(AC)\otimes\mathcal{A}(BC))\otimes SB,TC]\\ &\cong \textstyle\int_{C}[\mathcal{A}(AC)\otimes\int^{B} \mathcal{A}(BC)\otimes SB,TC]\\ &\cong \textstyle\int_{C}[\mathcal{A}(AC)\otimes SC,TC],\end{aligned}$$

$$\begin{aligned}(S\backslash T)A &= \textstyle\int_{C}[\int^{B} SB\otimes P(BAC),TC]\\ &\cong \textstyle\int_{C}[SC\otimes\mathcal{A}(AC),TC]\end{aligned}$$

for all $S,T \in [\mathcal{A},\mathcal{V}]$ and $A \in \mathcal{A}$.

It is easy to find instances of this biclosed structure and several commonly-occurring examples are given below.

Example 5.1 If A is a comonoid in $V_{\#}$ with only one object then its hom-object is a Hopf monoid in V, and $[P,V]$ is the usual biclosed category of modules over this monoid (cf. [3] IV §5).

Example 5.2 If V is cartesian closed then $V_{\#}$ is a cartesian monoidal category, hence every V-category A admits a unique (commutative) comonoid structure in $V_{\#}$ with the diagonal functor $A \to A{\times}A$ as comultiplication and the unique functor $A \to I$ as counit. Taking A small, the reduced tensor-product formula given above shows $[P,V]$ to be cartesian closed.

Example 5.3 Let $F\colon \mathcal{E}ns \to V$ be the monoidal closed functor described in Example 4.4, and let C be any $\mathcal{E}ns$-category. $\mathcal{E}ns$ is cartesian closed so C is a comonoid in $\mathcal{E}ns_{\#}$ and this induces an evident (commutative) comonoid structure on $F_{\#}C$. Hence, when C is small, the category $[F_{\#}C,V]$, whose underlying $\mathcal{E}ns$-category $[F_{\#}C,V]_0$ consists of the ordinary $\mathcal{E}ns$-functors from C to V_0 and the $\mathcal{E}ns$-natural transformations between them, automatically admits a symmetric monoidal closed structure over V. For $V = Ab$, this fact was pointed out by P. Freyd in [4].

The types of premonoidal category noted here, and in section 4, are far from being exhaustive. We have not, for instance, considered the premonoidal category which yields the following canonical biclosed structure on the category $[A^{op}\otimes A,V]$ of "bimodules" over an arbitrary small category A:

$$(S*T)(AB) = \int^C S(AC) \otimes T(CB) = S(AC) \underline{\otimes} T(CB),$$

$$(T/S)(AB) = \int_C [S(BC), T(AC)],$$

$$(S\backslash T)(AB) = \int_C [S(CA), T(CB)],$$

$$J(AB) = A(AB)$$

$$\ell^* = A(AC) \underline{\otimes} T(CB) \xrightarrow[y]{} T(AB),$$

$$r^* = T(AC) \underline{\otimes} A(CB) \xrightarrow[\underline{c}]{} A(CB) \underline{\otimes} T(AC) \xrightarrow[y]{} T(AB),$$

$$a^* = (R(AC) \underline{\otimes} S(CD)) \underline{\otimes} T(DB) \xrightarrow[\underline{a}]{} R(AC) \underline{\otimes} (S(CD) \underline{\otimes} T(DB)),$$

where $R,S,T \in [A^{op} \otimes A, V]$ and $(A,B) \in A^{op} \otimes A$ (axioms MC2 and MC3 for this definition of * are easily verified). In this case it is decidedly easier to describe the biclosed functor category than to give an explicit premonoidal structure on $A^{op} \otimes A$.

In conclusion I wish to express my gratitude to Professor Max Kelly for his helpful advice and discussions during the preparation of this article.

<u>Remark</u> The editor informs me that J. Bénabou, in lectures in 1967, had proposed the consideration of premonoidal categories, in the case $V = \mathcal{E}ns$, defining the functor P by considerations of section 4 above.

References

[1] Day, B.J. and Kelly, G.M., Enriched functor categories. Reports of the Midwest Category Seminar III (Lecture Notes in Mathematics, Volume 106, 1969), 178 - 191.

[2] Eilenberg, S. and Kelly, G.M., A generalisation of the functorial calculus, J. Algebra 3, 1966, 366 - 375.

[3] Eilenberg, S. and Kelly, G.M., Closed categories, Proc. Conf. on Cat. Alg. (La Jolla 1965), (Springer-Verlag 1966), 421 - 562.

[4] Freyd, P., Representations in abelian categories, Proc. Conf. on Cat. Alg. (La Jolla 1965), (Springer-Verlag 1966), 95 - 120.

[5] Kelly, G.M., Adjunction for enriched categories, Reports of the Midwest Category Seminar III (Lecture Notes in Mathematics, Volume 106, 1969), 166 - 177.

[6] Kock, A., On monads in symmetric monoidal closed categories, Aarhus Universitet, Preprint Series 1967/68, No. 14 (to appear in Archiv der Math.).

[7] Kock, A., Closed categories generated by commutative monads, Aarhus Universitet, Preprint Series 1968/69, No. 13 (to appear in J. Austr. Math. Soc.).

[8] Lambek, J., Deductive systems and categories, Category Theory, Homology Theory and their Applications 1 (Lecture Notes in Mathematics, Volume 86, 1969), 76 - 122.

[9] Linton, F.E.J., Autonomous equational categories, J. Math. Mech. 15, 1966, 637 - 642.

The University of New South Wales

RELATIONAL ALGEBRAS

Michael Barr

Received October 20, 1969

Introduction:

All notation not otherwise specified in this paper is taken from the introduction to the [ZTB]. Let $\underline{T}$ denote the category of topological spaces and continuous mappings, $\underline{C}$ be the full subcategory whose objects are the compact hausdorff spaces and $\underline{S}$ the category of sets. The obvious functor $U: \underline{C} \to \underline{S}$ has a left adjoint F which is best described by saying that FX is the set of all ultrafilters on X with the hull-kernel topology. If $\mathbb{T} = (T, \eta, \mu)$ is the triple coming from the adjoint pair, the natural functor $\underline{C} \to \underline{S}^{\mathbb{T}}$ is an equivalence. This associates to a compact hausdorff space C the pair (UC,c) where $c: TUC \to UC$ is given by $c(\underline{u}) = \lim \underline{u}$ for $\underline{u}$ an ultrafilter on UC.

In an arbitrary topological space V there is still given a relation $x: TUV \to UV$ which associates to an ultrafilter $\underline{u}$ the set $x(\underline{u})$ of all its limits (possibly empty). Moreover it is well known that this convergence relation determines the topology uniquely and that continuity of mappings can also be described by it. It thus seemed plausible that by a suitable axiomatization of the notion of relational algebra of a triple (or of a theory) it could be shown that $\underline{T}$ is the category of relational algebras of the theory whose algebras are $\underline{C}$. That this is so is the main result of this paper.

Section 1 tabulates a few properties of the category $\underline{R}$ of sets and relations. Section 2 includes the definitions of relational pre-algebras and establishes their basic properties. In section 3 the main theorem 3.1 about $\underline{T}$ is proved and in section 4 a few additional examples are given.

Everything here is done for triples over the category $\underline{S}$. Presumably much of this could be done over other categories, at least those in which there is a good notion of what are relations. Manes conjectures (and is attempting to prove) that the analog of (3.1) is true for any varietal category and its compact algebra triple (see [Ma], § 5).

In carrying out this work I had several stimulating discussions with Basil Rattray who independently proved (3.1). I would also like to thank the National Research Council of Canada for its support.

1. Relations.

Before talking about relational algebras we will here tabulate some properties of the category $\underline{R}$ of sets and relations. If $r: X \to Y$ is a relation it has a standard factorization

$$X \xrightarrow{dr^{-1}} \Gamma_r \xrightarrow{cr} Y$$

where $\Gamma_r \subset X \times Y$ is the graph of r and the functions cr, dr are the restrictions to Γ_r of the coordinate projections. We write

$r^{-1} = dr.cr^{-1}$. We call r epi, mono, e.d. (everywhere defined) or a p.f. (partial function) according as cr is epi, cr is mono, dr is epi or dr is mono, respectively. To define the composite of two relations it is only necessary to define a composition of the form $f^{-1}g$ where f and g are functions, for then we define $r.s = cr.dr^{-1}.cs.cs^{-1}$. Consider a commutative diagram:

$$\begin{array}{ccccc} & & \cdot & & \\ & {}^{v}\nearrow & & \searrow^{g} & \\ \cdot & & & & \cdot \\ & {}_{u}\searrow & & \nearrow_{f} & \\ & & \cdot & & \end{array} \tag{1.1}$$

where u, v, f, g are all functions. In general we have only $uv^{-1} \subset f^{-1}g$ with equality if (1.1) is a weak pullback. In particular we define $f^{-1}g$ to be uv^{-1} when (1.1) is a pullback.

It is well known that the category $\underline{R}$ is a 2-category, the hom-sets being partially ordered by $r \subset s$ in $\underline{R}(X,Y)$ if $\Gamma_r \subset \Gamma_s$. The following proposition is also well known (see e.g., [Mac]).

<u>Proposition 1.2</u>. For any relation $r\colon X \to Y$

i) r is epi iff $rr^{-1} \supset Y$

ii) r is mono iff $r^{-1}r \subset X$

iii) r is e.d. iff $r^{-1}r \supset X$

iv) r is a p.f. iff $rr^{-1} \subset Y$.

Of course combinations of these can be worked out as well. The most important characterizes r as a function iff $r^{-1}r \supset X$ and $rr^{-1} \subset Y$.

If T: $\underline{S} \to \underline{S}$ is a functor it induces a pseudofunctor, also denoted by T: $\underline{R} \to \underline{R}$ given by $T(cr.dr^{-1}) = Tcr.T(dr)^{-1}$. This means that for r·s composable relations, $T(r,s) \subset Tr.Ts$ and that $r \subset s \Rightarrow Tr \subset Ts$. Notice that $T(r^{-1}) = (Tr)^{-1}$ under this definition and will be written Tr^{-1}. If we have a diagram

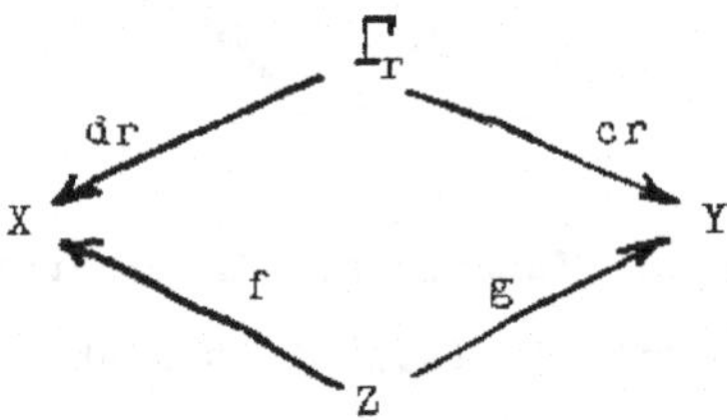

then $gf^{-1} \subset r$ iff there is a mapping $Z \to \Gamma_r$ making both triangles commute and $gf^{-1} = r$ iff that mapping is epi. In that case, since endofunctors on $\underline{S}$ preserve epimorphisms, it follows that the induced map $TZ \to T\Gamma_r$ is also epi. We have defined Tr to have the graph Γ_{Tr} which is the image of $T\Gamma_r$ in $TX \times TY$. The result is that the induced map $TZ \to \Gamma_{Tr}$ is still epi and so $Tr = Tg.Tf^{-1}$ is independent of the factorization. Then to see that T is a pseudofunctor it is sufficient to show that $T(f^{-1}g) \subset Tf^{-1}.Tg$, for then we have, after factoring $dr^{-1}.cs = p.q^{-1}$,

$$\begin{aligned} T(r.s) &= T(cr.dr^{-1}.cs.ds^{-1}) = T(cr.p.q^{-1}.ds^{-1}) \\ &= T(cr.p).T(q^{-1}.ds^{-1}) = Tcr.Tp.T((ds.q)^{-1}) \\ &= Tcr.Tp.(T(ds.q))^{-1} = Tcr.Tp.(Tds.Tq)^{-1} \\ &= Tcr.Tp.Tq^{-1}.Tds^{-1} = Tcr.T(pq^{-1}).Tds^{-1} \\ &= Tcr.T(dr^{-1}.cs).Tds^{-1} \subset Tcr.Tdr^{-1}.Tcs.Tds^{-1} \\ &= Tr.Ts \end{aligned}$$

Clearly equality holds if either r is a function or s is an inverse function. Now suppose that u and v are the pullback of f and g as in 1.1. After we apply T the diagram will still commute, but might fail to be a pullback. Thus, $T(f^{-1}g) = T(u.v^{-1}) = Tu.Tv^{-1} \subset Tf^{-1}.Tg$. If $\alpha: T \to T_1$ is a natural transformation of functors on S, it becomes pseudonatural on $\underline{R}$, which in this context means that $\alpha Y.Tr \subset T_1 r.\alpha X$.

2. Relational Algebras

We recall some of the definitions from the introduction to [ZTB]. A triple $\mathbb{T} = (T, \eta, \mu)$ on $\underline{X}$ is an endofunctor T of $\underline{X}$ together with natural transformations $\eta: T \to \underline{X}$ (= identity functor) and $\mu: T^2 \to T$ satisfying $\mu.T\mu = \mu.\mu T$, $\mu.T\eta = \mu.\eta T = T$. The category $\underline{X}^{\mathbb{T}}$ of $\mathbb{T}$-algebras has as objects (X,x) where X is an object and x: TX $\to$ X a morphism of $\underline{X}$ such that

$$x.Tx = x.\mu X$$

and

$$x.\eta X = X .$$

By a <u>relational</u> $\mathbb{T}$- <u>prealgebra</u> we mean a pair (X,x) where $x\colon TX \to X$ is a relation. A mapping $f\colon (X,x) \to (Y,y)$ of relational pre-algebras is a function $f\colon X \to Y$ such that $f.x \subset y.Tf$. The category of relational prealgebras and mappings of prealgebras is denoted by $\underline{S}^{P(\mathbb{T})}$. A relational pre-algebra (X,x) is called a relational algebra if

$$x.Tx \subset x.\mu X$$

and

$$x.\eta X \supset X .$$

This definition is a precise generalization of algebra because any inclusion between actual maps in $\underline{S}$ is an equality. We let $\underline{S}^{R(\mathbb{T})} \subset \underline{S}^{P(\mathbb{T})}$ denote the full subcategory consisting of relational algebras. That $\underline{S}^{\mathbb{T}} \subset \underline{S}^{R(\mathbb{T})} \subset \underline{S}^{P(\mathbb{T})}$ are full inclusions follows from the fact that an inclusion among functions is an equality. We will show that each is a coreflective inclusion (by which we mean it has a left adjoint). First we show that limits and colimits are easily computed in $\underline{S}^{P(\mathbb{T})}$.

<u>Proposition 2.1</u>. $\underline{S}^{P(\mathbb{T})}$ is complete and cocomplete.

<u>Proof</u>: Limits (= projective limits) are computed in $\underline{S}^{P(\mathbb{T})}$ exactly as in $\underline{S}^{\mathbb{T}}$. For example, if (X_i,x_i) is a family of relational

prealgebras their product in $\underline{S}^{P(\mathbb{T})}$ is $(X = \prod X_i, \prod x_i.a)$ where $a: T(\prod X_i) \to \prod TX_i$ is determined by $p_i.a = Tp_i$ and p_i is the ith coordinate projection. Equalizers are computed similarly. The same holds for relational algebras. As for colimits, first consider the same family (X_i, x_i). Let $b: \coprod TX_i \to T(\coprod X_i)$ be determined by $b.u_i = Tu_i$, where u_i denotes the ith coordinate injection. Then I claim that $(\coprod X_i, \coprod x_i.b^{-1})$ is the coproduct. First we must show the $u_i: X_i \to \coprod X_i$ are morphisms, i.e., that $u_i.x_i \subset \coprod x_i.b^{-1}.Tu_i$. Now $b.u_i = Tu_i$ (by 1.2.iii) $= b^{-1}.Tu_i$; then $u_i.x_i = \coprod x_i.u_i \subset \coprod x_i.b^{-1}.Tu_i$. (Note: we have been using u_i for different coproduct injections.) Now if $f_i: (X_i, x_i) \to (Y,y)$ is given for each i, then the f_i, extend to a unique $f: \coprod X_i \to Y$ such that $f.u_i = f_i$. Then $f.\coprod x_i.u_i = f.u_i.x_i = f_i.x_i \subset y.Tf_i =$ $= y.Tf.Tu_i = y.Tf.bu_i$ for all i and so by uniqueness of morphisms from a coproduct, $f.\coprod x_i \subset y.Tf.b$. Then $f.\coprod x_i.b^{-1} \subset y.Tf.b.b^{-1} \subset y.Tf$ (by 1.2.iv). Thus f is a morphism of relational prealgebras. Uniqueness is clear.

Now suppose $d^0, d^1: (X,x) \to (Y,y)$ are given. Let $d: Y \to Z$ be the coequalizer of the set morphisms d^0 and d^1 and $z = d.y.Td^{-1}: TZ \to Z$. Then $d.y \subset d.y.Td^{-1}.Td$ (by 1.2iii) $= z.Td$ and so d is a morphism. If $f: (Y,y) \to (W,w)$ coequalizes d^0 and d^1 then there is induced a unique $g: Z \to W$ with $g.d = f$. Moreover, $g.z = g.d.y.Td^{-1} = f.y.Td^{-1} \subset w.Tf.Td^{-1} = w.Tg.Td.Td^{-1} \subset w.Tg$ (by 1.2.iv).

<u>Remark</u>: Note that the limits and colimits preserve the underlying sets. This could have been predicted from the fact that the underlying set functor $\underline{S}^{P(\mathbb{T})} \to \underline{S}$ has both adjoints, namely $X \mapsto (X, \eta X^{-1})$ and $X \mapsto (X, X \times TX)$ being left and right adjoint, respectively. Each of these is a relational algebra so the same remarks apply to $\underline{S}^{R(\mathbb{T})} \to \underline{S}$.

<u>Proposition 2.2</u>. (X,x) is a relational algebra iff $x \supset \eta X^{-1}$ and $x.Tx.\mu X^{-1} \subset x$.

<u>Proof</u>: If (X,x) is a relational algebra then $X \subset x.\eta X$ so $\eta X^{-1} \subset x.\eta X.\eta X^{-1} \subset x$. Also $x.Tx.\mu X^{-1} \subset x.\eta X.\eta X^{-1} \subset x$. To see the converse we suppose that $x \supset \eta X^{-1}$; we have $x.\eta X \supset \eta X^{-1}.\eta X \supset X$. Similarly, if $x.Tx.\mu X^{-1} \subset x$, then $x.Tx \subset x.Tx.\mu X^{-1}.\mu X \subset x.\mu X$.

<u>Proposition 2.3</u>. The obvious functor $\underline{S}^{R(\mathbb{T})} \to \underline{S}^{P(\mathbb{T})}$ has a left adjoint.

<u>Proof</u>: Let $\eta = \eta X$ and $\mu = \mu X$. If (X,x) is a relational prealgebra, we define an ordinal sequence of relations $x_n: TX \to X$ as follows. Let $x_o = x \cup \eta^{-1}$. Having defined x_m for all $m < n$, define $x_n = \bigcup_{m<n} x_m$ if m is a limit ordinal and $x_n = x_{n-1}.Tx_{n-1}.\mu^{-1}$ otherwise. Clearly $x_o \supset x$ and $x_o \supset \eta^{-1}$. Assuming $x_m \supset x_k$ for all $k < m < n$, we have $x_n \supset x_k$ for all $k < n$ if n is a limit ordinal and otherwise

$$x_n = x_{n-1}.Tx_{n-1}.\mu^{-1} \supset x_{n-1}.T\eta^{-1}.\mu^{-1} =$$

$$= x_{n-1}.(\mu.T\eta)^{-1} = x_{n-1}.$$ Now since $TX \times X$ is a set this

ascending chain of subsets represented by the graphs of these x_n must eventually terminate, at which point $x_n = x_{n-1} = \bar{x}$ and so by 2.2 $(Tx, \bar{x})$ is a relational algebra. If $f: (X,x) \to (Y,y)$ is a morphism of (X,x) into the relational algebra (Y,y), then $f.x \subset y.Tf$. Also considering the diagram

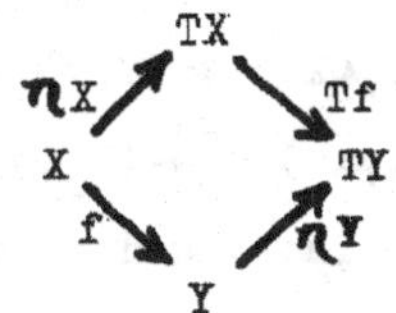

as a diagram like 1.1, we have $f.\eta^{-1} \subset \eta^{-1}.Tf \subset y.Tf$ so that $f.x_o \subset y.Tf$. Similarly if n is a limit ordinal and $f.x_m \subset y.Tf$ for all $m \subset n$ then certainly $f.x_n \subset y.Tf$. Finally, if $f.x_{n-1} \subset y.Tf$ then $f.x_n = f.x_{n-1}.Tx_{n-1}.\mu^{-1} \subset y.T(f.x_{n-1}).\mu^{-1}$ (since f is a function)

$\subset y.T(y.Tf).\eta^{-1} \subset y.Ty.T^2f.\mu^{-1} \subset y.Ty.\mu^{-1}.\mu.T^2f.\mu^{-1} \subset y.Tf.\mu.\mu^{-1} \subset y.Tf.$

Hence $f.\bar{x} \subset y.Tf$ from which the result easily follows.

<u>Proposition 2.4</u>. The obvious functor $\underline{S} \to \underline{S}^{R(\mathbb{T})}$ has a left adjoint.

<u>Proof</u>: This proof differs from the preceding in two respects. First the adjoint does not preserve the underlying set (by example). Second, the proof is nonconstructive, being an application of the adjoint functor theorem. It is clear from the description of limits in $\underline{S}^{P(\mathbb{T})}$ that the functor $\underline{S}^{\mathbb{T}} \to \underline{S}^{P(\mathbb{T})}$ preserves them. Also $\underline{S}^{R(\mathbb{T})}$ being a coreflective subcategory of $\underline{S}^{P(\mathbb{T})}$ computes its limits

in that category so that $\underline{S}^{\mathbb{T}} \to \underline{S}^{R(\mathbb{T})}$ preserves limits. So it is only necessary to find a solution set. Given (X,x), consider all algebras of the form $(TX,\mu X)/\sim$ for which the composite $X \xrightarrow{\eta X} TX \longrightarrow TX/\sim$ is a mapping of relational algebras. This is clearly small. If $f\colon (X,x) \to (Y,y)$ is a mapping of X into a $\mathbb{T}$-algebra, $y.Tf\colon TX \to Y$ is a function and $y.Tf\mu X = y.\mu Y.T^2f =$ $= y.T(y.Tf)$ so $y.Tf\colon (TX,\mu X) \to (Y,y)$ is a map of algebras. As usual, it factors as $(TX,\mu X) \xrightarrow{b} (Y_0,y_0) \xrightarrow{a} (Y,y)$ where (Y_0,y_0) is a factor algebra of $(TX,\mu X)$ and a is mono. Then
$a.b.\eta X = y.Tf.\eta X = y.\eta Y.f = f$ and
$a.b.\eta X.x = f.x \subset y.Tf = y.Ta.T(b.\eta X) = a.y_0.T(b.\eta X)$ and a is mono so $(b.\eta X).x \subset y_0.T(b.\eta X)$. Thus $b.\eta X\colon (X,x) \to (Y_0,y_0)$ factors f and the codomain is a member of the given set. We note in passing that this gives a new proof of the cocompleteness of $\underline{S}^{\mathbb{T}}$.

3. Topological spaces

Theorem 3.1. There is a natural equivalence J between the category $\underline{T}$ of topological spaces and continuous maps and the category $\underline{S}^{R(\mathbb{T})}$ where $\mathbb{T}=(T,\eta,\mu)$ is the compact hausdorff spaces triple.

The proof is given by a series of propositions. We let $\underline{u}$ denote an ultrafilter on X.

Proposition 3.2. If C is a topological space, $X = UC$ is its underlying set and $x\colon TX \to X$ is given by taking $x(\underline{u})$ to be the set of limits of $\underline{u}$, then $x.\eta \supset X$ and $x.Tx \subset x.\mu$.

Proof: The first statement is trivial, being merely the statement that if $\underline{p}$ is the principal ultrafilter at the point $p \in X$, then $p \in x(\underline{p})$ which is certainly true in any topological space. We can factor x as $TX \xleftarrow{j} DX \xleftarrow{d'x} \Gamma_x \xrightarrow{cx} X$ where j is mono, and d'x and cx are epi (the latter because of principal ultrafilters). T preserves both monos and epis so that

$$T^2X \xleftarrow{Tj} TDX \xleftarrow{Td'x} T\Gamma_x \xrightarrow{Tcx} TX$$

is the same type of decomposition. The mapping Tj consists of taking an ultrafilter on DX and using it as a filter base on TX. The filter generated is an ultrafilter. Thus if $\underline{a} \in T^2X$, $Tx(\underline{a}) \neq \emptyset$ iff given $A \in \underline{a}$, $DA = A \cap DX \in \underline{a}$ also. Now suppose that $\underline{a} \in T^2X$, $Tx(\underline{a}) = \underline{u}$ and $x(\underline{u}) = p$. Suppose that $\mu(\underline{a}) = \underline{v}$. Then we must show that $x(\underline{v}) = p$ as well. This means showing that every open neighborhood U of p is in $\underline{v}$ or, from the definition of μ, that $\{\underline{w} | U \in \underline{w}\} \in \underline{a}$. Now U is open which means that $x(\underline{w}) \cap U \neq \emptyset$. Suppose we had $A = \{\underline{w} | U \notin \underline{w}\} \in \underline{a}$. Now there is an ultrafilter $\underline{b}$ on $T\Gamma_x$ whose projections are $\underline{a}$ and $\underline{u}$ and there would have to be a $B \in \underline{b}$ whose projections were A on the one hand and some U_1 on the other. If $(\underline{w}, q) \in B$, then $U \notin \underline{w}$ but $q \in x(\underline{w})$. As noted above, $x(\underline{w}) \cap U = \emptyset$ and so $q \notin U$. Therefore, $U \cap U_1 = \emptyset$ which is a contradiction. Hence $A \notin \underline{a}$ and since $\underline{a}$ is an ultrafilter, its complement, $\{\underline{w} | U \in \underline{w}\} \in \underline{a}$. This completes the proof.

Proposition 3.3. If C and D are topological spaces with underlying sets X and Y and convergence mappings x and y respectively, then f: $X \to Y$ underlies a continuous mapping iff $f.x \subset x.Tf$.

Proof: This is nothing but a translation of a well known theorem which states that f is continuous iff whenever $\underline{u}$ is an ultrafilter on X and $\underline{u}$ converges to p then $Tf(\underline{u})$ converges to fp (see [E-G]).

Corollary 3.4. There is a natural J: $\underline{T} \to \underline{S}^{R(\mathbb{T})}$ which is full and faithful.

Proof: Of course J is defined by JC = (UC, convergence) as above and the preceding two propositions state that J is a well defined functor and is full. It is clearly faithful since U is.

Now suppose (X,x) is a relational algebra. We must show it is J of something. If e: $U \to X$ is a subset inclusion we define a new subset $\bar{e}$: $\bar{U} \to X$ by the following diagram

$$(3.5) \qquad \begin{array}{ccccc} TX & \xleftarrow{dx} & \Gamma_x & \xrightarrow{cx} & X \\ {\scriptstyle Te}\uparrow & & \uparrow{\scriptstyle g} & & \uparrow{\scriptstyle \bar{e}} \\ TU & \xleftarrow{dU} & \Gamma_U & \xrightarrow{cU} & \bar{U} \end{array}$$

in which Γ_U is the pullback of Te and dx and $\bar{e}.cU$ is the mono/epi factorization of $cx.\gamma$. In words, $\bar{U}$ is the set of "limits" of all ultrafilters containing U.

Proposition 3.6. $U \mapsto \bar{U}$ is a closure operator.

Proof: We must show a) $\bar{\emptyset} = \emptyset$, b) $U \subset \bar{U}$, c) $\overline{U_1 \cup U_2} = \bar{U}_1 \cup \bar{U}_2$ and d) $\bar{U} = \bar{\bar{U}}$. Part a) is trivial since $T\emptyset = \emptyset$ and a pullback along $\emptyset$ is empty. b) is also easy for $p \in U$; the principal ultrafilter $\underline{p}$ contains U and converges to p. This implies that $\overline{U_1 \cup U_2} \supset \bar{U}_1 \cup \bar{U}_2$

so showing c) requires only the reverse inclusion. But if an ultrafilter $\underline{u}$ contains $U_1 \cup U_2$ then it must contain one of them and so we get the reverse inclusion. d) is harder. Again, we already have one inclusion. $\overline{U} \subset \overline{\overline{U}}$. It is easily seen to be sufficient to find a relation $t: \overline{\overline{U}} \to \overline{U}$ such that $\overline{\overline{e}} \subset \overline{e}.t$. For then $\text{image}(\overline{\overline{e}}) \subset \text{image}(\overline{e}.t) \subset \text{image}(\overline{e})$ and the result is proved. We define $t = cU.dU^{-1}.\ \mu U.dU.TcU^{-1}.d\overline{U}.c\overline{U}^{-1}$. In words, take a point of $\overline{\overline{U}}$ and find an ultrafilter on $\overline{U}$ converging to this. Represent that in turn by an ultrafilter on TU. Converge that under μ to an ultrafilter on U and converge that under x. Now recalling that $g.dU^{-1} = dx^{-1}.Te$ since Γ_U is a pullback and $f.f^{-1} \supset$ identity when f is epi, we have:

$$\begin{aligned}
\overline{e}.t &= \overline{e}.cU.dU^{-1}.\mu U.TdU.TcU^{-1}.d\overline{U}.c\overline{U}^{-1} \\
&= cx.g.dU^{-1}.\mu U.TdU.TcU^{-1}.d\overline{U}.c\overline{U}^{-1} \\
&= cx.dx^{-1}.Te.\mu U.TdU.TcU^{-1}.d\overline{U}.c\overline{U}^{-1} \\
&= x.\mu X.T^2e.TdU.TcU^{-1}.d\overline{U}.c\overline{U}^{-1} \\
&\supset x.Tx.T^2e.TdU.TcU^{-1}.d\overline{U}.c\overline{U}^{-1} \\
&= x.Tcx.Tdx^{-1}.T^2e.TdU.TcU^{-1}.d\overline{U}.c\overline{U}^{-1} \\
&= x.Tcx.Tg.TdU^{-1}.TdU.TcU^{-1}.d\overline{U}.c\overline{U}^{-1} \\
&\supset x.T\overline{e}.TcU.TcU^{-1}.d\overline{U}.c\overline{U}^{-1} \\
&= x.T\overline{e}.d\overline{U}.c\overline{U}^{-1} \\
&= cx.dx^{-1}.dx.\overline{g}.c\overline{U}^{-1} \\
&\supset cx.\overline{g}.c\overline{U}^{-1} = \overline{\overline{e}}.c\overline{U}.c\overline{U}^{-1} = \overline{\overline{e}}
\end{aligned}$$

It is well known that such a closure operator induces a unique topology on X. There is now only one thing left to complete the proof of Theorem 3.1.

<u>Proposition 3.7</u>. Suppose $(X,x) \in \underline{S}^{R(\mathbb{T})}$ and the closure is defined as above. Then for any $\underline{u} \in TX$, $p \in x(\underline{u})$ iff $p \in \bigcap \{\overline{U} \mid U \in \underline{a}\}$.

<u>Remark</u>. Since this latter is the formula for ultrafilter convergence under a closure operator, this means that x is convergence (as described in the introduction) under that topology.

<u>Proof</u>: One way is clear. For if $p \in x(\underline{u})$ and $U \in \underline{u}$, $\underline{u}$ is an ultrafilter containing U and p is one of its limits under x so $p \in \overline{U}$. To go the other way we suppose that $U \in \underline{u}$ converges to $p \in \overline{U}$. We want to show that $p \in x(\underline{u})$. The idea of the proof is to find an $\underline{a} \in T^2X$ with $\mu(\underline{a}) = \underline{u}$ while $Tx(\underline{a}) = \underline{p}$, the principal ultrafilter at p. Then $p \in x(\underline{p}) \subset x.\mu(\underline{a}) = x(\underline{u})$. We begin by constructing an ultrafilter $\underline{b}$ on Γ_x. Let $B = \{(\underline{v},p) \mid p \in x(\underline{v})\} \subset \Gamma_x$ and for each $U \in \underline{u}$ let $B_U = \{(\underline{v},q) \mid U \in \underline{v}$ and $q \in x(\underline{v})\}$. Then for $U_1, U_2 \in U$, $B_{U_1} \cap B_{U_2} \supset B_{U_1 \cap U_2} \neq \emptyset$ since $U_1 \cap U_2 \neq \emptyset$. Moreover, since $p \in \overline{U}$ for each $U \in \underline{u}$, we can find a $\underline{v}$ such that $U \in \underline{U}$ and $p \in x(\underline{v})$. Then $(\underline{v},p) \in B \cap B_U$. Thus B together with $\{B_U\}$ generates a proper filter which is contained in an ultrafilter $\underline{b}$ on Γ_x. Clearly one projection is $\underline{p}$ since B projects to $\{p\}$. The other projection is $\underline{a}$ and it is clear that for all $U \in \underline{u}$, $\{\underline{v} \mid U \in v\} \supset dx(B_U)$ and thus is in $\underline{a}$. Hence $U \in \mu(\underline{a})$ for all $U \in \underline{u}$ and so $\underline{u} = \mu(\underline{a})$.

4. Other examples.

In this section we consider a few other examples of relational algebras although in no case are the results as striking as the preceding. The simplest example is the identity triple $\mathbb{I} = (I,1,1)$. Here the laws simply reduce to $x: X \twoheadrightarrow X$ subject to $x.x \subset x$ and $x \supset X$. Thus an algebra for this triple is a set with a transitive, reflexive relation, usually called a preordered set. It is simple to check that mappings are exactly the order preserving functions. It is perhaps interesting to note in this connection that the algebras, $\underline{\underline{S}}$, are a coreflexive (hence tripleable) subcategory of preordered sets.

The next example is the triple $\mathbb{T} = (T,\eta,\mu)$ where $TX = X+1$, where + denotes the coproduct. The category of algebras is the category of pointed sets. A relational algebra is a set X with a relation $x: X+1 \twoheadrightarrow X$. Now X+1 is a coproduct and so x is determined by its restrictions $x|X$ and $x|1$. The first restriction is a preorder, as before, and the second is just a subset $X_0 \subset X$. The unitary law is automatic (as soon as $x|X$ is a preorder) while the associative law requires that X_0 be a ray. That is, if $p \in X_0$, $q \leq p$ (or $q \in x(p)$) then $q \in X_0$ also. Thus the algebras are pairs $(X, X_0 \subset X)$ where X is a preordered set and X_0 is a ray. Morphisms are maps of pairs which preserve the order as well.

Similar considerations apply to the category of relational models of the triple where TX is the set of ultrafilters on X+1.

The algebras are pairs (X,X_o) where X is a topological space and $X_o \subset X$ is a closed subspace. Mappings are continuous maps of pairs.

No other interesting examples are known to us. It seems fairly clear that Cat, for example, is a full subcategory of the relational models of the theory of monoids (however the nullary operation 1 becomes the set of objects of a category!) but just being a full subcategory is not very informative. (E.g. the recent work of the "Čech school" seems to show that practically everything is fully embedded in the category of semigroups).

References

[ZTB] - B. Eckmann, ed., "Seminar on Triples and Algebraic Homology Theory", Lecture Notes in Mathematics, no. 80, Springer Verlag, 1969.

[Ma] - E. Manes, A triple theoretic construction of compact algebras, ZTB, 91-118.

[E-G] - R. Ellis and W. H. Gottschalk, Homomorphisms of transformation groups, Trans. Amer. Math. Soc. 94 (1960), 258-271.

[Mac] - S. Mac Lane, An algebra of additive relations, Proc. NAS-USA 47, (1961), 1043-1051.

ITERATED COTRIPLES

by

H. Applegate and M. Tierney[(1)]

Received October 27, 1969

0. INTRODUCTION

In this paper we will be concerned with dissecting a given adjoint pair

$$\underline{A} \underset{s}{\overset{r}{\rightleftarrows}} B\ (\varepsilon, \eta): r \dashv s$$

into progressively "smaller" adjoint pairs by, essentially, iterating the formation of the cotriple $\mathbb{G} = (rs, \varepsilon, r\eta s)$ generated by $r \dashv s$.

The notation s and r is meant to suggest the singular and realization functors of the model induced adjoint pairs defined in [1], since these are the ones we are most interested in for applications. In this case, $\underline{B} = (\underline{M}^*, S)$ for some small category $\underline{M}$ called the category of models - S being the category of sets - and $r \dashv s$ is canonically induced by a given functor $I: \underline{M} \to \underline{A}$ (the definition is given in 3(ii)). Here the cotriple $\mathbb{G} = (rs, \varepsilon, r\eta s)$ is called the model induced cotriple, or the

(1) The second named author was partially supported by the NSF under Grant GP-8618.

<u>density cotriple</u> by some other authors. The term "singular" and "realization" refer to the case $\underline{M} = \underline{\Delta}$ - the simplicial category - and $\underline{A} = \underline{\text{Top}}$, where s and r are the standard singular and geometric realization functors of topology.

In §1 we consider the method of dissection itself, and develop a mechanism that takes an adjoint pair and associates to it an ordinal tower of adjoint pairs having the given one as its 0th coordinate. The basis for this construction consists of two lemmas, which provide us, respectively, with the inductive step of the procedure at either a successor or a limit ordinal.

Having constructed the tower, we consider its limit $\hat{\underline{A}}$ in §2. It turns out, if $\underline{B}$ is well powered, that the limit adjoint pair

$$\hat{\underline{A}} \underset{\hat{s}}{\overset{\hat{r}}{\leftrightarrows}} \underline{B}$$

is a coreflection. As a result, we can factor the original pair in the form

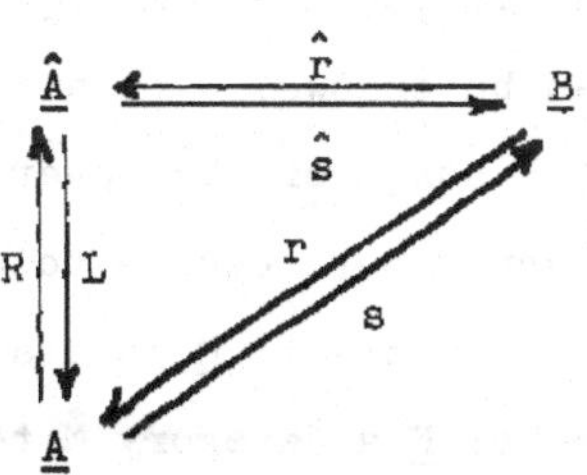

where $(\hat{\varepsilon},\hat{\eta})\colon \hat{r} \dashv \hat{s}$, $\hat{\varepsilon}\colon \hat{r}\hat{s} \xrightarrow{\approx} I$, $L \dashv R$, $L\hat{r} = r$, and L reflects isomorphisms.

The above factorization is exploited in §3(i) to prove results about the existence of adjoints to projections into categories of fractions. Specifically, let $\underline{A} \underset{s}{\overset{r}{\rightleftarrows}} \underline{B}$ be a given adjoint pair and write Σ for the class of morphisms of $\underline{B}$ made invertible by r. Then if

$$r_{\Sigma} : \underline{B} \longrightarrow \underline{B}[\Sigma^{-1}]$$

denotes the canonical projection of $\underline{B}$ into its category of fractions with respect to Σ, we show that r_{Σ} has an adjoint iff there exists a factorization of the above type for $r \dashv s$. §3(ii) consists of an application of §2 to the theory of [1]. In particular, we consider a model induced adjoint pair

$$\underline{A} \underset{s}{\overset{r}{\rightleftarrows}} (\underline{M}^*, S)$$

and show that the $\underline{M}$-objects of $\underline{A}$, defined in [1], are precisely those objects of $\underline{A}$ that persist to the limit $\hat{\underline{A}}$. That is, they are those objects of $\underline{A}$ that can be endowed with a coalgebra structure, in a compatible way, at each stage of the tower associated to $r \dashv s$. In §3(iii) we apply our results to obtain Lambek completions of small categories, where if $\underline{M}$ is a small category we mean by a Lambek completion of $\underline{M}$ a category $\hat{\underline{M}}$ together with an embedding I: $\underline{M} \to \hat{\underline{M}}$ such that $\hat{\underline{M}}$ is complete and cocomplete, and I is full, faithful, dense, cogenerating, continuous and cocontinuous.

(The terms are explained in §3(iii).) In fact, what we prove is the following. Let I: $\underline{M} \to \underline{A}$ be full, faithful, and codense with $\underline{A}$ complete. Then $\underline{A}$ is cocomplete and we can build the tower associated to the model induced adjoint pair. Denoting its limit by $\hat{\underline{A}}$, we show that I can be lifted to

$$\hat{I}: \underline{M} \longrightarrow \hat{\underline{A}} ,$$

and $\hat{\underline{A}}$ together with $\hat{I}$ forms a Lambek completion of $\underline{M}$. As an example of such a functor I we could take the contravariant Yoneda embedding

$$Y^*: \underline{M} \longrightarrow (\underline{M},S)^* .$$

We might remark here that these techniques have recently been extended by Dubuc to the case of closed categories and strong functors to obtain corresponding strong completions.

1. THE MACHINE.

We prove first the two lemmas alluded to in § 0. Namely, start with an adjoint pair

$$\underline{A} \underset{s}{\overset{r}{\rightleftarrows}} \underline{B} \quad (\varepsilon,\eta): r \dashv s$$

and let $\mathbf{G} = (rs, \varepsilon, r\eta s)$ be the cotriple it generates in $\underline{A}$. Forming the category $\underline{A}_{\mathbf{G}}$ of $\mathbf{G}$-coalgebras we obtain the standard diagram

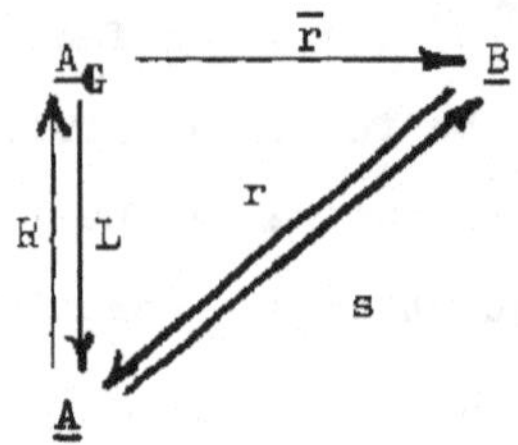

where the Eilenberg-Moore comparison $\bar{r}$ is given by

$$\bar{r}B = (rB,\ r\eta B)\ ,$$

and $L \dashv R$ by

$$L(A,\theta) = A$$
$$RA = (GA,\ \delta A) = \bar{r}sA.$$

Then the first lemma, due originally to Jon Beck, is the following - we append a sketch of the proof since none is available in the literature.

<u>1.1 Lemma.</u>

(a) $\bar{r}$ has an adjoint $\bar{s}$: $\underline{A}_{\mathbb{G}} \to \underline{B}$ iff for each $(A,\theta) \in \underline{A}_{\mathbb{G}}$, the diagram

$$sA \underset{\eta sA}{\overset{s\theta}{\rightrightarrows}} srsA$$

has an equalizer in $\underline{B}$ - this equalizer, if it exists, is $\bar{s}(A,\theta)$.

(b) If $\bar{r}$ has an adjoint $\bar{s}$ and $(\bar{\varepsilon},\bar{\eta})$: $\bar{r} \dashv \bar{s}$, then

$$\bar{\varepsilon}(A,\theta):\ \bar{r}\bar{s}(A,\theta) \longrightarrow (A,\theta)$$

is an isomorphism iff r preserves the equalizer

$$\bar{s}(A,\Theta) \xrightarrow{j(A,\Theta)} sA \overset{s\Theta}{\underset{\eta sA}{\rightrightarrows}} srsA .$$

(c) If $\bar{r}$ has an adjoint $\bar{s}$, and $\bar{\varepsilon}: \bar{r}\bar{s} \longrightarrow 1$ is an equivalence, then

$$\bar{\eta}: 1 \longrightarrow \bar{s}\bar{r}$$

is an equivalence iff r reflects isomorphisms.

<u>Proof</u>:

(a) Assuming $\bar{s}$ exists with $(\bar{\varepsilon},\bar{\eta}): \bar{r} \dashv \bar{s}$, we let

$$j: \bar{s} \longrightarrow sL$$

be the natural transformation corresponding under $r \dashv s$ to

$$L\bar{\varepsilon}: L\bar{r}\bar{s} = r\bar{s} \longrightarrow L .$$

Thus, for a coalgebra (A,Θ), $j(A,\Theta)$ is the composite

$$\bar{s}(A,\Theta) \xrightarrow{\eta\bar{s}(A,\Theta)} sr\bar{s}(A,\Theta) \xrightarrow{sL\bar{\varepsilon}(A,\Theta)} sA ,$$

and a simple argument shows that

$$(*) \quad \bar{s}(A,\Theta) \xrightarrow{j} sA \overset{s\Theta}{\underset{\eta sA}{\rightrightarrows}} srsA$$

is an equalizer in $\underline{B}$. (To simplify notation, we often write just j for $j(A,\Theta)$.) Conversely, assume that the equalizer (*) exists in $\underline{B}$ for all $(A,\Theta) \in \underline{A}_{\mathbb{G}}$, thereby defining a functor $\bar{s}: \underline{A}_{\mathbb{G}} \longrightarrow \underline{B}$

and a natural transformation $j\colon \bar{s} \to sL$ To define $\bar{\varepsilon}\colon \bar{r}\bar{s} \to 1$ and $\bar{\eta}\colon 1 \to \bar{s}\bar{r}$, recall that an $\underline{A}$-morphism $\Theta\colon A \to GA$ is a coalgebra structure iff

$$A \xrightarrow{\Theta} GA \underset{\delta A}{\overset{G\Theta}{\rightrightarrows}} G^2 A$$

is an equalizer in $\underline{A}$. Since rj equalizes $G\Theta$ and δA - apply r to (*) - rj factors uniquely through Θ as

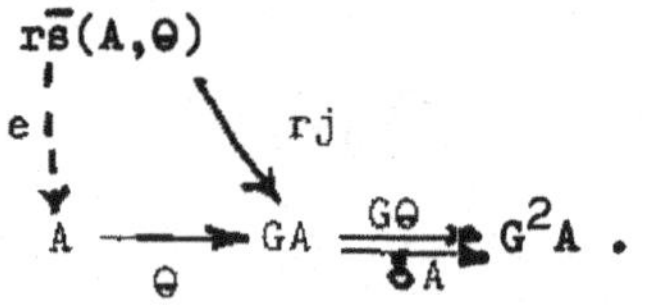

$$(**)\qquad \begin{array}{ccccc} \bar{r}\bar{s}(A,\Theta) & & & & \\ {\scriptstyle e}\downarrow & \searrow^{rj} & & & \\ A & \xrightarrow[\Theta]{} & GA & \underset{\delta A}{\overset{G\Theta}{\rightrightarrows}} & G^2 A\ . \end{array}$$

In fact, $e = \varepsilon A \cdot rj$ and it is easy to see that e is the underlying of a coalgebra morphism

$$\bar{\varepsilon}(A,\Theta)\colon \bar{r}\bar{s}\,(A,\Theta) \longrightarrow (A,\Theta)\ .$$

$\bar{\eta}B\colon B \to \bar{s}\bar{r}B$ is defined by the diagram

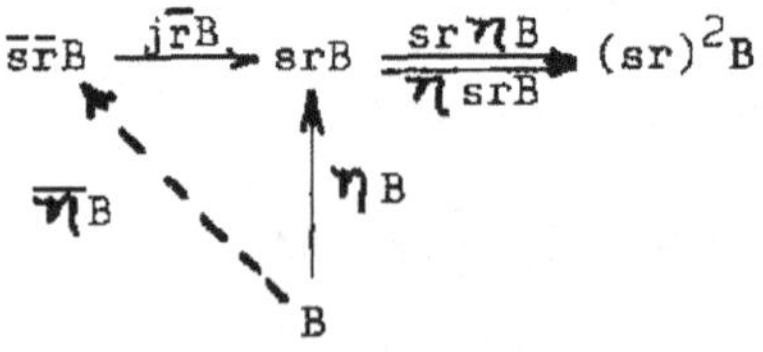

$$(***)\qquad \begin{array}{ccccc} \bar{s}\bar{r}B & \xrightarrow{j\bar{r}B} & srB & \underset{\eta srB}{\overset{sr\eta B}{\rightrightarrows}} & (sr)^2 B \\ & {\scriptstyle \bar{\eta}B}\nwarrow & \uparrow{\scriptstyle \eta B} & & \\ & & B & & \end{array},$$

for ηB always equalizes $sr\eta B$ and ηsrB. The verification that $(\bar{\varepsilon},\bar{\eta})\colon \bar{r} \dashv \bar{s}$ is left to the reader.

(b) This is immediate, since in (**) rj is an equalizer iff $e = L\mathcal{E}(A,\Theta)$ is an isomorphism iff $\mathcal{E}(A,\Theta)$ is.

(c) If $(\bar{\varepsilon},\bar{\eta})\colon \bar{r} \dashv \bar{s}$ and $\bar{\varepsilon}\colon \bar{r}\bar{s} \xrightarrow{\sim} 1$, then by (b) and (***) we see that both ηB and $j\bar{r}B$ become equalizers upon application of r. It follows that $r\bar{\eta}B$ is an isomorphism, making $\bar{\eta}B$ one if r reflects isomorphisms. The converse is evident since L reflects isomorphisms.

Now if we assume $\underline{B}$ has equalizers, then by iterating the construction of 1.1(a) we can build a tower of adjoint pairs indexed by the non-negative integers having the given one at stage 0. However, using only this technique we run out of steam at the first limit ordinal ω, so for this step we need another lemma.

Therefore, let $\underline{B}$ be an arbitrary category and denote by $(\underline{B}, \text{Cat})$ the category of categories under $\underline{B}$. The objects of $(\underline{B}, \text{Cat})$ are functors $\underline{B} \xrightarrow{r} \underline{A}$, and morphisms are commutative triangles

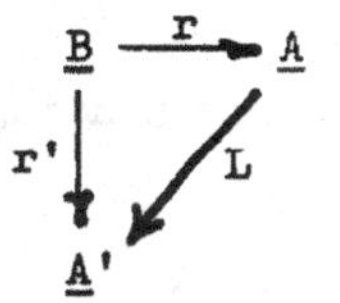

Let $\text{Ad}(\underline{B}, \text{Cat})$ be the full subcategory of $(\underline{B}, \text{Cat})$ consisting of functors $\underline{B} \xrightarrow{r} \underline{A}$ for which there exists an adjoint $s\colon \underline{A} \to \underline{B}$.

<u>1.2 Lemma</u>.

If $\underline{B}$ is complete, so is $\text{Ad}(\underline{B}, \text{Cat})$.

<u>Proof</u>: Let I be a small index category with objects α and morphisms $f\colon \alpha \to \beta$, and let

$$I \to \mathrm{Ad}(\underline{B}, \mathrm{Cat})$$

be a functor, i.e., we have a functor $\underline{B} \xrightarrow{r_\alpha} \underline{A}_\alpha$ for each $\alpha \in I$ and compatible diagrams

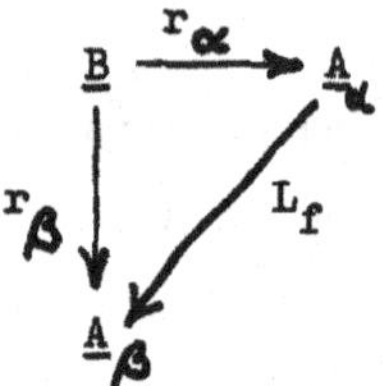

for each $f\colon \alpha \to \beta$. Now for each $\alpha \in I$ choose an adjoint $s_\alpha : \underline{A}_\alpha \to \underline{B}$ to r_α, and let $\eta_\alpha : 1 \to s_\alpha r_\alpha$, $\varepsilon_\alpha : r_\alpha s_\alpha \to 1$ be the unit and counit, respectively, of the adjointness $r_\alpha \dashv s_\alpha$. If $f\colon \alpha \to \beta$, let

$$j_f\colon s_\alpha \to s_\alpha L_f$$

be the natural transformation corresponding under $r_\beta \dashv s_\beta$ to

$$L_f \varepsilon_\alpha : r_\beta s_\alpha = L_f r_\alpha s_\alpha \to L_f ,$$

i.e.,

$$j_f = s_\beta L_f \varepsilon_\alpha \cdot \eta_\beta s_\alpha .$$

If $\alpha \xrightarrow{f} \beta \xrightarrow{g} \gamma$, then one readily establishes the commutativity of the following diagrams:

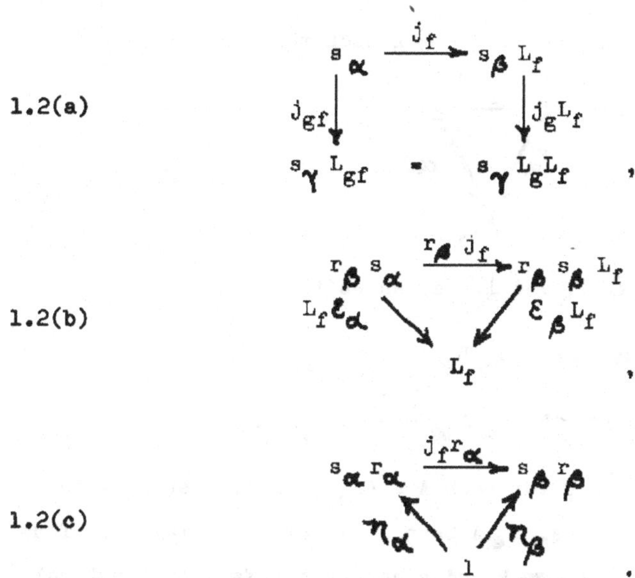

This done, let

$$\underline{A} = \varprojlim_{\alpha} \underline{A}_{\alpha}$$

with projections $L_{\alpha} : \underline{A} \longrightarrow \underline{A}_{\alpha}$. Recall that objects $A \in \underline{A}$ are families $A = (A_{\alpha})$ of objects $A_{\alpha} \in \underline{A}_{\alpha}$ indexed by $\alpha \in I$ such that if $f: \alpha \longrightarrow \beta$ then $L_f A_{\alpha} = A_{\beta}$. Similarly, morphisms

$$g: A = (A_{\alpha}) \longrightarrow (A'_{\alpha}) = A'$$

are families $g = (g_{\alpha})$ where $g_{\alpha} : A_{\alpha} \longrightarrow A'_{\alpha}$ and $L_f g_{\alpha} = g_{\beta}$ if $f: \alpha \longrightarrow \beta$. The projections are given by

$$L_{\alpha} A = A_{\alpha}, \; L_{\alpha} g = g_{\alpha} .$$

Now the r_α define a unique $r\colon \underline{B} \to \underline{A}$ such that

$$\begin{array}{ccc} \underline{B} & \xrightarrow{r} & \underline{A} \\ & {\scriptstyle r_\alpha}\searrow \quad \swarrow{\scriptstyle L_\alpha} & \\ & \underline{A}_\alpha & \end{array}$$

commutes for each $\alpha \in I$, and 1.2(a) shows that

$$\alpha \rightsquigarrow s_\alpha L_\alpha A = s_\alpha A_\alpha$$
$$(\alpha \xrightarrow{f} \beta) \rightsquigarrow (s_\alpha A_\alpha \xrightarrow{j_f A_\alpha} s_\beta A_\beta)$$

defines an I-diagram in $\underline{B}$ for each $A = (A_\alpha)$ in $\underline{A}$. Let sA with projections $j_\alpha A\colon sA \to s_\alpha L_\alpha A$ be a limit of this diagram. Thus we obtain a functor $s\colon \underline{A} \to \underline{B}$ and a compatible family of natural transformations $j_\alpha : s \to s_\alpha L_\alpha$.

It is clear that $r \dashv s$, for if $A = (A_\alpha) \in \underline{A}$ and $B \in \underline{B}$, then

$$\underline{A}(rB, A) = \varprojlim_\alpha \underline{A}_\alpha (r_\alpha B, A_\alpha) \approx \varprojlim_\alpha \underline{B}(B, s_\alpha A_\alpha)$$
$$\approx \underline{B}\,(B, \varprojlim_\alpha s_\alpha A_\alpha) = \underline{B}\,(B, sA)\ .$$

The unit and counit of this adjointness are easily determined. In fact, $\eta : 1 \to sr$ is characterized by the diagrams

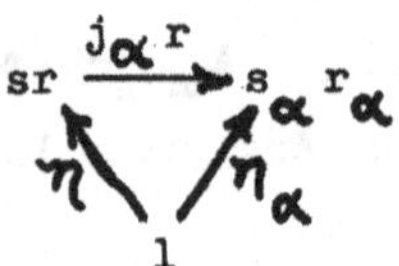

since the $\eta_\alpha : 1 \to s_\alpha r_\alpha$ are a compatible family by 1.2(c). $\varepsilon : rs \to 1$ is given by specifying

$$L_\alpha(\varepsilon) : r_\alpha s \longrightarrow L_\alpha$$

to be the transformation corresponding under $r_\alpha \dashv s_\alpha$ to $j_\alpha : s \to s_\alpha L_\alpha$, i.e.,

$$\begin{array}{ccc} r_\alpha s & \xrightarrow{r_\alpha j_\alpha} & r_\alpha s_\alpha L_\alpha \\ & {\scriptstyle L_\alpha(\varepsilon)} \searrow \quad \swarrow {\scriptstyle \varepsilon_\alpha L_\alpha} & \\ & L_\alpha & \end{array}$$

is defined to be commutative. An obvious verification shows that $\underline{B} \xrightarrow{r} \underline{A}$ is in fact a limit in Ad($\underline{B}$, Cat) of the original diagram.

We remark that 1.2 can be formulated in such a way as to be a generalization of 1.1(a), although we leave all details concerning this to the reader.

Now let $\underline{\Omega}$ be the category of ordinal numbers; i.e., the objects of $\underline{\Omega}$ are ordinal numbers and there is a unique morphism $\alpha \to \beta$ iff $\alpha \leq \beta$ in the usual ordering. Then using 1.1, 1.2, and transfinite induction we arrive at

<u>1.3 Theorem.</u>

Let $\underline{A}_o \underset{s_o}{\overset{r_o}{\rightleftarrows}} \underline{B}$ be an adjoint pair with $(\varepsilon_o, \eta_o) : r_o \dashv s_o$. Then if $\underline{B}$ is complete there is an $\underline{\Omega}^*$ diagram

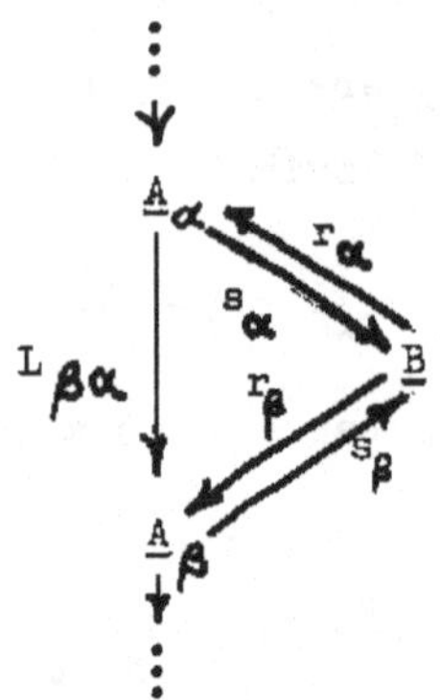

in Ad($\underline{B}$, Cat) with the following properties:

(a) For $\alpha = 0$ the adjoint pair is the given one;

(b) If $\alpha = \beta + 1$ is a successor, then

$$\underline{A}_\alpha = (\underline{A}_\beta)_{\mathbb{G}_\beta}$$

where $\mathbb{G}_\beta$ is the cotriple in $\underline{A}_\beta$ generated by $r_\beta \dashv s_\beta$, r_α is the Eilenberg-Moore comparison, and $L_{\beta\alpha}: \underline{A}_\alpha \to \underline{A}_\beta$ is the canonical underlying for coalgebras;

(c) If α is a limit ordinal, then

$$(\underline{B} \xrightarrow{r_\alpha} \underline{A}_\alpha) = \varprojlim_{\beta<\alpha} (\underline{B} \xrightarrow{r_\beta} \underline{A}_\beta)$$

with projection $L_{\beta\alpha}: \underline{A}_\alpha \to \underline{A}_\beta$ as in 1.2 with $I = \{\beta < \alpha\}^*$.

For the rest of this section we shall be concerned with developing various useful properties of this $\underline{\Omega}^*$ diagram. The diagram itself, to which all notation refers, should be considered as given and fixed in the following series of propositions.

First, as in 1.2, if $\beta < \alpha$ let $j_{\beta\alpha} : s_\alpha \to s_\beta L_{\beta\alpha}$ correspond to $L_{\beta\alpha}(\mathcal{E}_\alpha) : L_{\beta\alpha} r_\alpha s_\alpha = r_\beta s_\alpha \to L_{\beta\alpha}$ under $r_\beta \dashv s_\beta$. Then the diagrams corresponding to 1.2(a), 1.2(b), and 1.2(c) also hold for the $j_{\beta\alpha}$ and will be used without comment.

1.4 Proposition.

Each $j_{\beta\alpha}$ is monic.

Proof: We prove first by transfinite induction that for each α, $j_{o\alpha} : s_\alpha \to s_o L_{o\alpha}$ is monic. In fact, suppose this is true for all $\gamma < \alpha$.

If $\alpha = \gamma + 1$ is a successor, we have

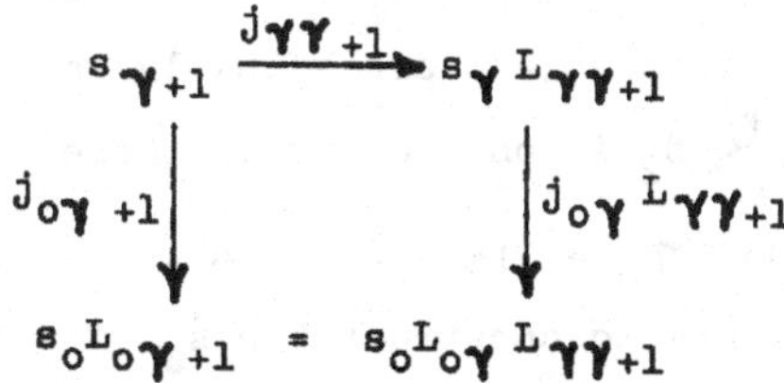

making $j_{o\alpha}$ monic since $j_{\gamma\gamma+1}$ is an equalizer. If α is a limit ordinal, then the $j_{\gamma\alpha} : s_\alpha \to s_\gamma L_{\gamma\alpha}$ for $\gamma < \alpha$ are the projections of a limit diagram. Thus again the diagram

$$\begin{array}{ccc} s_\alpha & \xrightarrow{j_{\gamma\alpha}} & s_\gamma L_{\gamma\alpha} \\ {\scriptstyle j_{o\alpha}}\downarrow & & \downarrow {\scriptstyle j_{o\gamma} L_{\gamma\alpha}} \\ s_o L_o \gamma & = & s_o L_{o\gamma} L_{\gamma\alpha} \end{array}$$

shows that $j_{o\alpha}$ is monic. Now for an arbitrary pair $\beta < \alpha$, we have

$$\begin{array}{ccc} s_\alpha & \xrightarrow{j_{\beta\alpha}} & s_\beta L_{\beta\alpha} \\ \Big\downarrow j_{o\alpha} & & \Big\downarrow j_{o\beta} L_{\beta\alpha} \\ s_o L_{o\alpha} & = & s_o L_{o\beta} L_{\beta\alpha} \end{array}$$

so that $j_{\beta\alpha}$ is monic since $j_{o\alpha}$ is.

Note that 1.4 shows the assumption of completeness in 1.3 to be unnecessarily strong. Examining the construction one sees that only reflexive equalizers and intersections of linearly ordered chains of subobjects of an object are actually used. (A reflexive equalizer $B_o \xrightarrow{e} B_1 \overset{f}{\underset{g}{\rightrightarrows}} B_2$ is one for which there exists $\eta : B_2 \longrightarrow B_1$ such that $\eta f = \eta g = 1$.) In fact, by 1.1(a) we need reflexive equalizers to construct the s_α for α a successor, and if α is a limit ordinal, then for $A_\alpha = (A_\gamma)_{\gamma<\alpha} \in \underline{A}_\alpha = \varprojlim_{\gamma<\alpha} \underline{A}_\gamma$ we have

$$s_\alpha A_\alpha = \varprojlim_{\gamma<\alpha} s_\gamma A_\gamma ,$$

where the morphisms in the diagram are the $j_{\gamma\gamma'} A_\gamma : s_\gamma A_\gamma \longrightarrow s_{\gamma'} A_{\gamma'}$ for $\gamma' < \gamma$. But by 1.4 each of these is monic, and each $s_\gamma A_\gamma$ represents a subobject of $s_o A_o$. $s_\alpha A_\alpha$, therefore, is simply the intersection of the $s_\gamma A_\gamma$. We shall refrain from weakening the hypothesis, however, since "complete" is easier to say than "has reflexive equalizers and intersections of linearly ordered chains of subobjects," and in examples it is generally the case that if the latter is true so is the former.

<u>1.5 Proposition.</u>

Each $L_{\beta\alpha}$ is faithful and reflects isomorphisms.

<u>Proof</u>: As in 1.4, we show first by transfinite induction that the proposition holds for each $L_{o\alpha}$. Thus suppose $L_{o\gamma}$ is faithful and reflects isomorphisms for $\gamma < \alpha$. If $\alpha = \gamma + 1$ we have

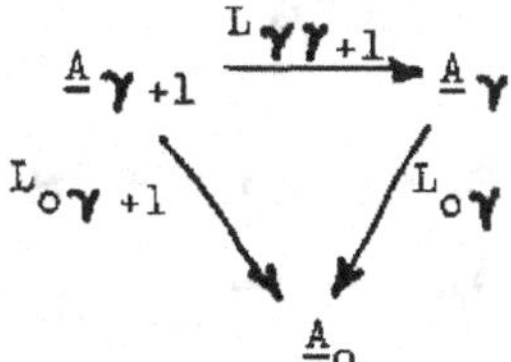

which gives the result. If α is a limit ordinal, then the $L_{\gamma\alpha}: \underline{A}_{\alpha} \longrightarrow \underline{A}_{\gamma}$ for $\gamma < \alpha$ are the projections of a limit, and

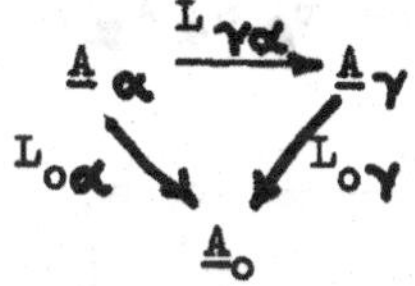

establishes the claim. Now for an arbitrary pair $\beta < \alpha$

$$\begin{array}{ccc} \underline{A}_{\alpha} & \xrightarrow{L_{\beta\alpha}} & \underline{A}_{\beta} \\ & {\scriptstyle L_{o\alpha}}\searrow \quad \swarrow{\scriptstyle L_{o\beta}} & \\ & \underline{A}_{o} & \end{array}$$

gives the general case.

We want to show, finally, that each $L_{\beta\alpha}$ has an adjoint, but for this we require a lemma.

<u>1.6 Lemma</u>.

For any ordinal α, let $(A_\alpha, \theta_\alpha) \in \underline{A}_{\alpha+1}$ and if $\beta < \alpha$ put $(A_\beta, \theta_\beta) = L_{\beta+1\,\alpha+1}(A_\alpha, \theta_\alpha)$. Then

$$\begin{array}{ccc} A_\beta & \xrightarrow{L_{\beta\alpha}\theta_\alpha} & r_\beta s_\alpha A_\alpha \\ & {\scriptstyle\theta_\beta}\searrow \quad \swarrow{\scriptstyle r_\beta j_{\beta\alpha} A_\alpha} & \\ & r_\beta s_\beta A_\beta & \end{array}$$

commutes.

<u>Proof</u>: Well, $\theta_\alpha : A_\alpha \longrightarrow r_\alpha s_\alpha A_\alpha$, and $A_\alpha = L_{\alpha\alpha+1}(A_\alpha, \theta_\alpha)$ so

$$L_{\beta+1\alpha}\theta_\alpha : (A_\beta, \theta_\beta) \longrightarrow L_{\beta+1\alpha} r_\alpha s_\alpha A_\alpha .$$

But

$$L_{\beta+1\alpha} r_\alpha s_\alpha A_\alpha = r_{\beta+1} s_\alpha A_\alpha = (r_\beta s_\alpha A_\alpha, r_\beta \eta_\beta s_\alpha A_\alpha)$$

so

$$L_{\beta+1\alpha}\theta_\alpha : (A_\beta, \theta_\beta) \longrightarrow (r_\beta s_\alpha A_\alpha, r_\beta \eta_\beta s_\alpha A_\alpha),$$

i.e.,

$$\begin{array}{ccc} A & \xrightarrow{L_{\beta\alpha}\theta_\alpha} & r_\beta s_\alpha A_\alpha \\ {\scriptstyle\theta_\beta}\downarrow & & \downarrow{\scriptstyle r_\beta \eta_\beta s_\alpha A_\alpha} \\ r_\beta s_\beta A_\beta & \xrightarrow[r_\beta s_\beta L_{\beta\alpha}\theta_\alpha]{} & r_\beta s_\beta r_\beta s_\alpha A_\alpha \end{array}$$

commutes. Now recall that

$$j_{\beta\alpha} A_\alpha : s_\alpha A_\alpha \longrightarrow s_\beta A_\beta$$

corresponds to

$$L_{\beta\alpha}(\mathcal{E}_\alpha A_\alpha): r_\beta s_\alpha A_\alpha \longrightarrow A_\beta$$

under $r_\beta \dashv s_\beta$, giving

$$\begin{array}{ccc} s_\alpha A_\alpha & \xrightarrow{\eta_\beta s_\alpha A_\alpha} & s_\beta r_\beta s_\alpha A_\alpha \\ & {\scriptstyle j_{\beta\alpha} A_\alpha} \searrow \quad \swarrow {\scriptstyle s_\beta L_{\beta\alpha} \mathcal{E}_\alpha A_\alpha} & \\ & s_\beta A_\beta & \end{array} .$$

Applying r_β to this and composing with the previous diagram gives the result, since $\mathcal{E}_\alpha A_\alpha \cdot \theta_\alpha = 1$.

1.7 Proposition.

$L_{\beta\alpha} \dashv r_\alpha s_\beta$ for $\alpha > \beta$.

Proof: For $\alpha > \beta$ we must define natural transformations

$$\mathcal{E}_{\beta\alpha}: L_{\beta\alpha} r_\alpha s_\beta = r_\beta s_\beta \longrightarrow 1$$
$$\eta_{\beta\alpha}: 1 \longrightarrow r_\alpha s_\beta L_{\beta\alpha}$$

satisfying the usual identities. The definition of $\mathcal{E}_{\beta\alpha}$ is obvious: $\mathcal{E}_{\beta\alpha} = \mathcal{E}_\beta$. To define $\eta_{\beta\alpha} A_\alpha: A_\alpha \longrightarrow r_\alpha s_\beta A_\beta$ where $A_\alpha \in \underline{A}_\alpha$ and $A_\beta = L_{\beta\alpha} A_\alpha$, we must consider two cases.

Case (1). $\alpha = \gamma+1$ is a successor. Then $A_\alpha = A_{\gamma+1} = (A_\gamma, \theta_\gamma)$ where $\theta_\gamma: A_\gamma \longrightarrow G_\gamma A_\gamma$ is a $\mathbb{G}_\gamma$ - coalgebra structure. However, θ_γ is also a morphism of coalgebras

$$\theta_\gamma : A_{\gamma+1} = (A_\gamma, \theta_\gamma) \longrightarrow (G_\gamma A_\gamma, \delta_\gamma A_\gamma) = r_{\gamma+1} s_\gamma A_\gamma,$$

and we let $\eta_{\beta\alpha} A_\alpha$ be the composite

$$A_{\gamma+1} \xrightarrow{\Theta_\gamma} r_{\gamma+1}s_\gamma A_\gamma \xrightarrow{r_{\gamma+1}j_{\beta\gamma}A_\gamma} r_{\gamma+1}s_\beta A_\beta \, .$$

Case (2). α is a limit ordinal. Then $A_\alpha = (A_\gamma)_{\gamma<\alpha}$ and $r_\alpha s_\beta A_\beta = (r_\gamma s_\beta A_\beta)_{\gamma<\alpha}$.

We define a compatible family

$$f_\gamma : A_\gamma \longrightarrow r_\gamma s_\beta A_\beta$$

as follows: $A_{\gamma+1} = (A_\gamma, \Theta_\gamma)$, and

$$f_\gamma = \begin{cases} A_\gamma \xrightarrow{\Theta_\gamma} r_\gamma s_\gamma A_\gamma \xrightarrow{r_\gamma j_{\beta\gamma} A_\gamma} r_\gamma s_\beta A_\beta & \text{if } \gamma > \beta \\ L_{\gamma\beta}\Theta_\beta & \text{if } \gamma \leq \beta \end{cases}$$

1.6 shows immediately that (f_γ) is a compatible family so we get

$$\eta_{\beta\alpha}A_\alpha = (f_\gamma) \quad .$$

Naturality in both cases is evident. In the special case that $A_\alpha = r_\alpha B$ for some $B \in \underline{B}$ we have

$$L_{\gamma\alpha}\eta_{\beta\alpha}r_\alpha B = r_\gamma j_{\beta\gamma}r_\gamma B \cdot r_\gamma \eta_\gamma B = r_\gamma \eta_\beta B$$

for $\gamma > \beta$, and this makes it easy to verify the required identities.

2. BEHAVIOR IN THE LIMIT.

Let

$$\underline{A}_0 \underset{s_0}{\overset{r_0}{\rightleftarrows}} \underline{B}\ (\varepsilon_0, \eta_0) : r_0 \dashv s_0$$

be an adjoint pair with $\underline{B}$ complete. Consider the diagram

$$\underline{\Omega}^* \longrightarrow \text{Ad}\ (\underline{B}, \text{Cat})$$

given by 1.3 for Ω the class of all ordinals. Although this is no longer a small diagram, let us try to proceed as in 1.2. So, first form the possibly illegitimate (i.e., non-locally small) category

$$\underline{A} = \varprojlim_{\alpha} \underline{A}_{\alpha} \ ,$$

and denote the projections of the limit by $L_\alpha : \underline{A} \longrightarrow \underline{A}_\alpha$. Then the $r_\alpha : \underline{B} \longrightarrow \underline{A}_\alpha$ define a functor $r: \underline{B} \longrightarrow \underline{A}$. $A = (A_\alpha) \in \underline{A}$ gives an $\underline{\Omega}^*$ diagram in $\underline{B}$ by $\beta < \alpha \leadsto j_{\beta\alpha} : s_\alpha A_\alpha \longrightarrow s_\beta A_\beta$, and if this diagram has a limit in $\underline{B}$, we can put

$$sA = \varprojlim_{\alpha} s_\alpha A_\alpha \ ,$$

yielding $r \dashv s$ as in 1.2. By 1.4 each $j_{\beta\alpha}$ is monic, so that the diagram we are considering is a descending ordinal chain of sub-objects of $s_0 A_0$. Thus, at this point let us assume that in addition to being complete, $\underline{B}$ is well-powered. If $B \in \underline{B}$ let $P(B)$ denote the set of equivalence classes of subobjects of B, and if $B_0 \longrightarrow B$ is monic, denote its equivalence class by $\bar{B}_0$. Then by the above, for each $(A_\alpha) \in \underline{A}$ we have an order reversing function

$$\Omega \longrightarrow P(s_o A_o)$$

given by

$$\alpha \rightsquigarrow \overline{s_\alpha A_\alpha} \quad .$$

What we need now is the following lemma, whose proof we add only because it may be unfamiliar to some category theorists.

2.1. Lemma.

Let X be a partially ordered set and $\alpha \rightsquigarrow x_\alpha$ an order reversing function $\Omega \longrightarrow X$. Then there exists an α such that for all $\beta \geq \alpha$, $x_\beta = x_\alpha$.

Proof: Suppose not. Then for all α there exists $\beta > \alpha$ such that $x_\beta < x_\alpha$. Define $f: \Omega \longrightarrow \Omega$ by

$$f(\alpha) = \inf\{\beta > \alpha \mid x_\beta < x_\alpha\} \ .$$

Then $f(\alpha) > \alpha$ and $x_{f(\alpha)} < x_\alpha$. Now define $g: \Omega \longrightarrow \Omega$ inductively by $g(o) = o$ and if $\beta > o$, $g(\beta) = f(\sup_{\alpha<\beta} g(\alpha))$. First of all, g is order preserving. In fact, if $\beta_1 > \beta_2$ then $g(\beta_1) = f(\sup_{\alpha<\beta_1} g(\alpha)) > \sup_{\alpha<\beta_1} g(\alpha) \geq g(\beta_2)$. Note that this gives $g(\alpha+1) = f\,g(\alpha)$. Also, the function $\alpha \rightsquigarrow x_{g(\alpha)}$ is injective, for if $\beta > \alpha$ then $\beta \geq \alpha+1 > \alpha$ and $x_{g(\beta)} \leq x_{g(\alpha+1)} = x_{fg(\alpha)} < x_{g(\alpha)}$. This is clearly absurd, since if we denote by $\bar{\bar{Y}}$ the cardinality of a set Y we can choose an ordinal β such that $\bar{\bar{\beta}} > \bar{\bar{X}}$. But by the above, the function $\alpha \rightsquigarrow x_{g(\alpha)}: \{\alpha \mid \alpha < \beta\} \longrightarrow X$ is injective giving $\bar{\bar{\beta}} \leq \bar{\bar{X}}$.

Going back to the original argument, we see by 2.1 that for each $A = (A_\alpha) \in \underline{A}$ the function

$$\Omega \longrightarrow P(s_o A_o)$$

must stabilize. That is, there is a β such that for all $\alpha \geq \beta$

$$\overline{s_\alpha A_\alpha} = \overline{s_\beta A_\beta} \quad .$$

or, equivalently,

$$j_{\beta\alpha} A_\alpha : s_\alpha A_\alpha \xrightarrow{\sim} s_\beta A_\beta \quad .$$

As a result, not only does

$$sA = \varprojlim_{\alpha} s_\alpha A_\alpha$$

exist in $\underline{B}$, but moreover the projections

$$j_\alpha : sA \longrightarrow s_\alpha A_\alpha$$

are isomorphisms for $\alpha \geq \beta$. Thus, we can define $\eta : 1 \rightarrow sr$ $\varepsilon : rs \rightarrow 1$ and as in §1 and we obtain $(\varepsilon, \eta) : r \dashv s$. Let us investigate ε more closely. First of all, because

$$\begin{array}{ccc} \underline{A} & \xrightarrow{L_\alpha} & \underline{A}_\alpha \\ & {\scriptstyle L_o} \searrow \quad \swarrow {\scriptstyle L_{o\alpha}} & \\ & \underline{A}_o & \end{array}$$

commutes for each α it follows that L_o is faithful and reflects isomorphisms since this is true of each $L_{0\alpha}$ (1.5). (Note that this makes $\underline{A}$ legitimate.) On the other hand, this fact plus the above diagram shows that every L_α is faithful and reflects isomorphisms. Therefore, if $A = (A_\alpha) \in \underline{A}$ then $\varepsilon A: rsA \longrightarrow A$ will be an isomorphism iff for some α, $L_\alpha(\varepsilon A): r_\alpha sA \longrightarrow A_\alpha$ is. Now recall that $L_\alpha(\varepsilon A)$ is the morphism corresponding under $r_\alpha \dashv s_\alpha$ to the projection $j_\alpha: sA \longrightarrow s_\alpha A_\alpha$, i.e.,

$$\begin{array}{ccc} r_\alpha sA & \xrightarrow{r_\alpha j_\alpha} & r_\alpha s_\alpha A_\alpha \\ & {\scriptstyle L_\alpha(\varepsilon A)} \searrow \quad \swarrow {\scriptstyle \varepsilon_\alpha A_\alpha} & \\ & A_\alpha & \end{array}$$

commutes. Choosing β so that for all $\alpha \geq \beta$ j_α is an isomorphism gives

$$j_{\beta+1}: sA \xrightarrow{\sim} s_{\beta+1}A_{\beta+1},$$

and also

$$j_{\beta\beta+1}: s_{\beta+1}A_{\beta+1} \xrightarrow{\sim} s_\beta A_\beta .$$

But if $A_{\beta+1} = (A_\beta, \theta_\beta)$, we know from 1.1(b) that

$$\varepsilon_{\beta+1}A_{\beta+1}: r_{\beta+1}s_{\beta+1}A_{\beta+1} \xrightarrow{\sim} A_{\beta+1}$$

iff r_β preserves the equalizer

$$s_{\beta+1}A_{\beta+1} \xrightarrow{j_{\beta\beta+1}} s_\beta A_\beta \underset{\eta_\beta s_\beta A_\beta}{\overset{s_\beta\theta_\beta}{\rightrightarrows}} s_\beta r_\beta s_\beta A_\beta ,$$

which it must since $j_{\beta\beta+1}$ is an isomorphism. Thus, we have finally

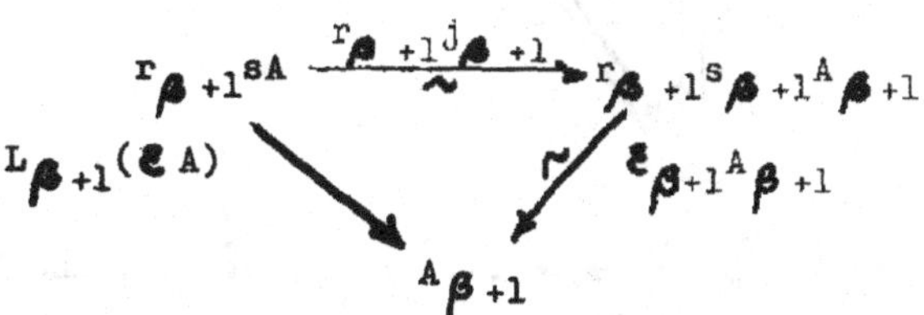

so that $L_{\beta+1}(\varepsilon A)$, and hence also εA, is an isomorphism.

For later use we remark that in the diagram

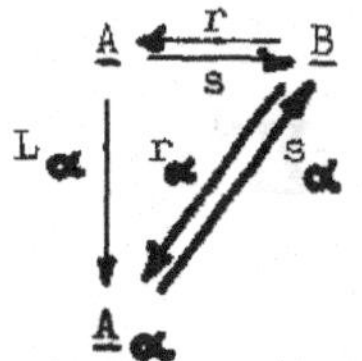

we can obtain, as in 1.7, $R_\alpha : \underline{A}_\alpha \longrightarrow \underline{A}$ so that $L_\alpha \dashv R_\alpha$ by setting $R_\alpha = rs_\alpha$.

3. APPLICATIONS

(i) Categories of fractions.

We have seen in § 2 that an adjoint pair

$$\underline{A} \underset{s}{\overset{r}{\rightleftarrows}} \underline{B} \ (\varepsilon, \eta): r \dashv s$$

with $\underline{B}$, say, complete and well-powered can be factored in the form

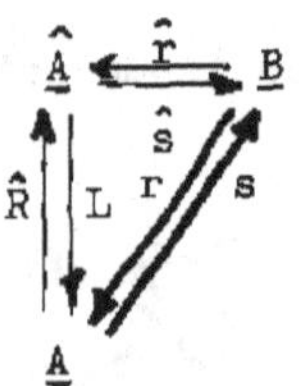

where $(\hat{\varepsilon},\hat{\eta}): \hat{r} \dashv \hat{s}$, $\hat{\varepsilon}: \hat{r}\hat{s} \xrightarrow{\approx} 1$, $L \dashv \hat{R}$, $L\hat{r} = r$, and L reflects isomorphisms. Let us see what this means for categories of fractions, assuming the reader is familiar with [2] Chapter 1.

Well, first let $\hat{\Sigma}$ be the class of all morphisms $\sigma \in \underline{B}$ made invertible by $\hat{r}$, and denote by

$$r_{\hat{\Sigma}} : \underline{B} \longrightarrow \underline{B}\,[\hat{\Sigma}^{-1}]$$

the canonical projection of $\underline{B}$ to the category of fractions of $\underline{B}$ with respect to $\hat{\Sigma}$. Then, since $\hat{\varepsilon}: \hat{r}\hat{s} \xrightarrow{\approx} 1$ we know by 1.3 of [2] that the canonical functor

$$\hat{H}: \underline{B}[\hat{\Sigma}^{-1}] \longrightarrow \underline{A}$$

such that

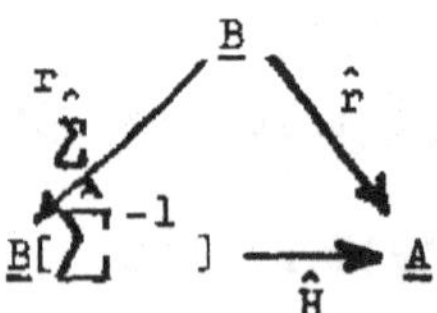

commutes is an equivalence of categories. Now let Σ be the class of morphisms $\sigma \in \underline{B}$ made invertible by r. Since $L\hat{r} = r$, $\hat{\Sigma} \subset \Sigma$, and since L reflects isomorphisms, $\Sigma \subset \hat{\Sigma}$. Thus $\hat{\Sigma} = \Sigma$ and we can replace the above diagram by

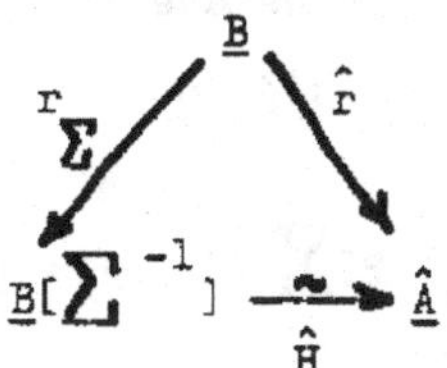

and the original factorization by

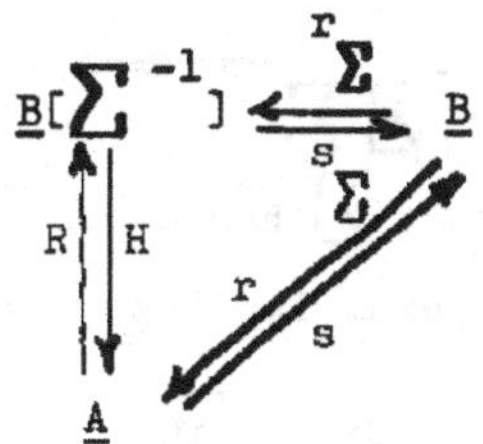

,

where $s_{\Sigma} = \hat{s}\hat{H}$, $\hat{H}R = \hat{R}$, and $H = L\hat{H}$ is the canonical functor. Thus, the original factorization, when it exists, is unique, and we have proved

3.1 Theorem.

For any adjoint pair $\underline{A} \underset{s}{\overset{r}{\rightleftarrows}} \underline{B}$ (ε, η): $r \dashv s$, if $\underline{B}$ is complete and well-powered then

$$r_{\Sigma} : \underline{B} \longrightarrow \underline{B}[\Sigma^{-1}]$$

has an adjoint.

Note that the hypothesis of 3.1 are quite different from those that one would impose if one were trying to prove this result by means of the adjoint functor theorem.

On the other hand, let us assume, for the given adjoint pair of 3.1, that r_{Σ} has an adjoint

$$s_{\Sigma} : \underline{B}[\Sigma^{-1}] \longrightarrow \underline{B} .$$

(We make no other assumptions on $\underline{B}$ here). Then it follows easily from 1.3 of [2] that $\varepsilon_{\Sigma} : r_{\Sigma} s_{\Sigma} \longrightarrow 1$ is an isomorphism, and with this, one can prove that Σ has a calculus of left fractions [2]. The latter fact, in turn, implies that in the canonical diagram

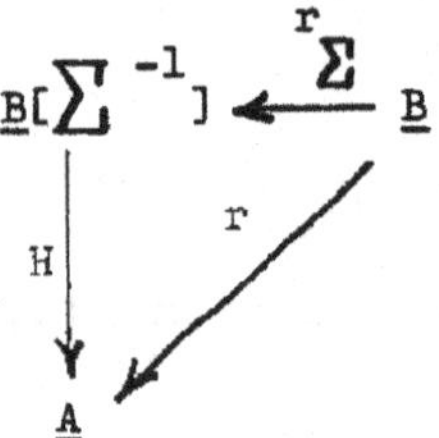

H reflects isomorphisms. Now let us prove the following lemma.

3.2 Lemma.

In the diagram

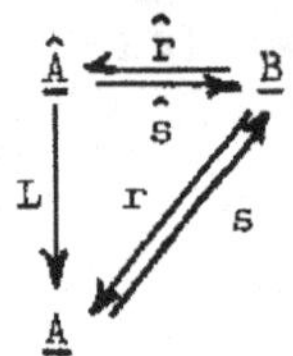

assume $(\varepsilon, \eta): r \dashv s$, $(\hat{\varepsilon}, \hat{\eta}): \hat{r} \dashv \hat{s}$, $\hat{\varepsilon}: \hat{r}\hat{s} \xrightarrow{\sim} 1$, and $L\hat{r} = r$. Then if $R = \hat{r}s$, we have $L \dashv R$.

Proof: Define $\bar{\varepsilon}: LR \to 1$ by setting $\bar{\varepsilon} = \varepsilon$ - since $LR = L\hat{r}s = rs$. Let $\bar{\eta}: 1 \to RL$ be the composite

$$1 \xrightarrow{\hat{\varepsilon}^{-1}} \hat{r}\hat{s} \xrightarrow{\hat{r}j} \hat{r}sL,$$

where $j: \hat{s} \to sL$ is determined by the diagram

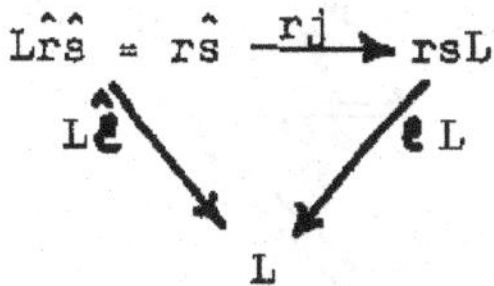

i.e., j corresponds under $r \dashv s$ to $L\hat{\varepsilon}$. By definition, the composite

$$L \xrightarrow{L\bar{\eta}} LRL \xrightarrow{\bar{\varepsilon}L} L$$

is the identity of L. Consider the composite

$$R \xrightarrow{\bar{\eta}R} RLR \xrightarrow{R\bar{\varepsilon}} R,$$

i.e., the composite

$$\hat{r}s \xrightarrow{\hat{\varepsilon}\hat{r}s^{-1}} \hat{r}\hat{s}\hat{r}\hat{s} \xrightarrow{\hat{r}j\hat{r}s} \hat{r}s\hat{r}s \xrightarrow{\hat{r}s\varepsilon} \hat{r}s.$$

Since

$$\hat{r}s \xrightarrow{\hat{r}\hat{\eta}s} \hat{r}\hat{s}\hat{r}s \xrightarrow{\hat{\varepsilon}\hat{r}s} \hat{r}s$$

is the identity of $\hat{r}s$, it follows that $\hat{\varepsilon}\hat{r}s^{-1} = \hat{r}\hat{\eta}s$, and

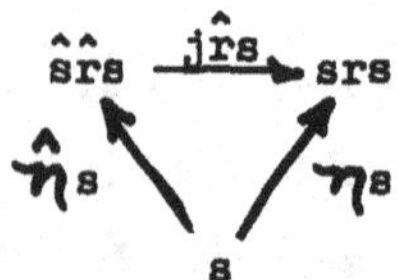

commutes. But this gives the result, since $r \dashv s$.

Note that this gives an alternative proof of the fact that $R_\alpha = rs_\alpha$ is adjoint to the $L_\alpha : \underline{A} \to \underline{A}_\alpha$ of § 2.

Applying 3.2 to the diagram

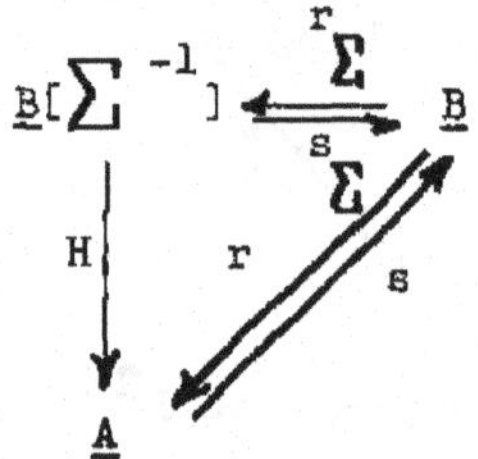

we see that $R = r_\Sigma s$ is adjoint to H. This, the existence of an adjoint to r_Σ is equivalent to the existence of a factorization of the type used in 3.1.

(ii) M-objects.

Let

$$\underline{A}_0 \underset{s_0}{\overset{r_0}{\rightleftarrows}} \underline{B}\ (\varepsilon_0, \eta_0) : r_0 \dashv s_0$$

be an adjoint pair with $\underline{B}$, say, complete so that the ordinal tower of § 1 exists. Using the notation of § 1, we say $A_0 \in \underline{A}_0$ is

<u>special</u> iff there is a monomorphism

$$k_o : B \longrightarrow s_o A_o$$

such that the composite

$$r_o B \xrightarrow{r_o k_o} r_o s_o A_o \xrightarrow{\varepsilon_o A_o} A_o$$

is an isomorphism. We denote this composite by e_o, and we will in general use the letter e to refer to adjoints of morphisms such as k_o. We say A_o <u>persists</u> iff there is an $A \in \underleftarrow{\lim}\ \underline{A}_\alpha$ such that $L_o A = A_o$.

Suppose A_o is special. Then it is easy to see that

$$\theta_o = r_o k_o \cdot e_o^{-1}$$

is the unique $\mathbb{G}_o$ - coalgebra structure for A_o such that

$$e_1 : r_1 B \xrightarrow{\approx} A_1$$

where $L_{o1}\ e_1 = e_o$ and $A_1 = (A_o, \theta_o)$. $\mathbb{G}_o$, of course, is the cotriple $(r_o s_o, \varepsilon_o, r_o \eta_o s_o)$. Let

$$k_1 = s_1 e_1 \cdot \eta_1 B .$$

Then the diagram

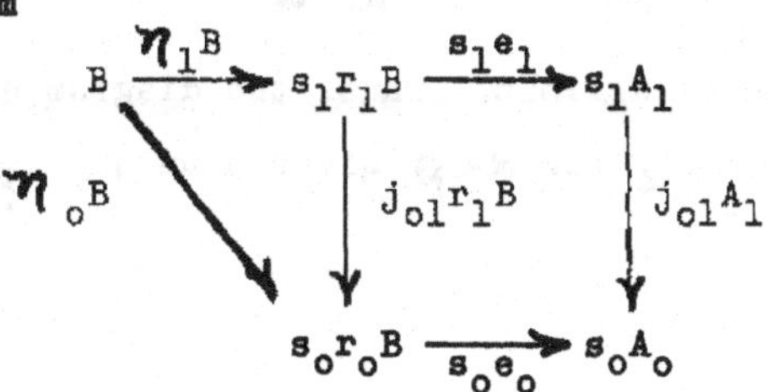

shows that

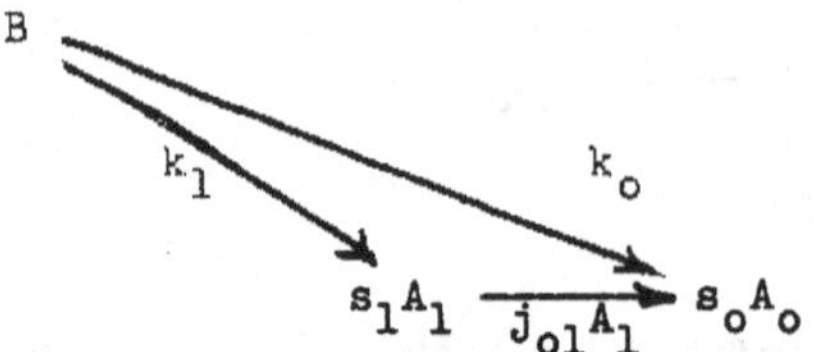

commutes. Now k_1 is monic since k_o is, and its adjoint is e_1 which is an isomorphism. Thus we have advanced one stage. For the general inductive step assume that we have defined, for all $\alpha < \beta$, $A_\alpha \in \underline{A}_\alpha$ and a monomorphism $k_\alpha : B \rightarrow s_\alpha A_\alpha$ such that e_α is an isomorphism, and if $\alpha' \leq \alpha$ then $L_{\alpha'\alpha}(e_\alpha) = e_{\alpha'}$ and

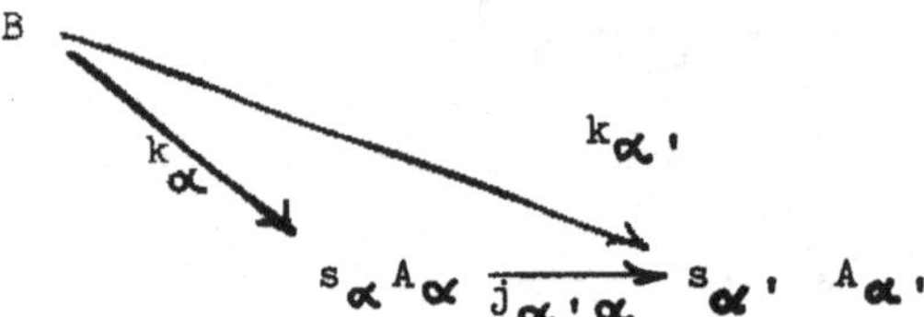

commutes. If $\beta = \alpha+1$ is a successor we proceed as above to obtain k_β. Suppose β is a limit ordinal. Then
$A_\beta = (A_\alpha)_{\alpha<\beta} \in \underline{A}_\beta$ and

$$s_\beta A_\beta \xrightarrow{j_{\alpha\beta}} s_\alpha A_\alpha$$

are the projections of a limit. Thus, the diagram expressing the compatibility of the k_α for $\alpha<\beta$ gives a unique k_β such that

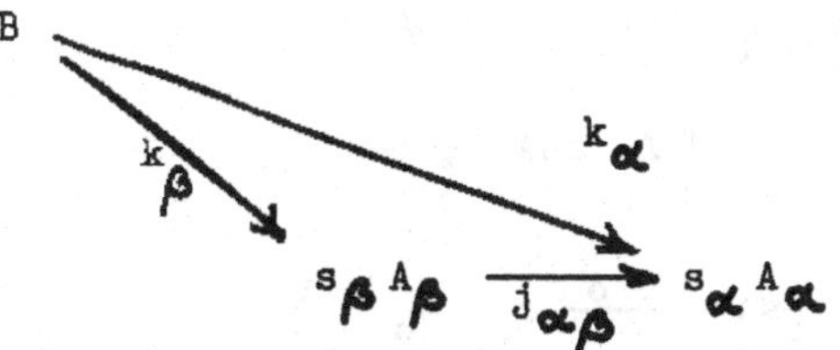

commutes for $\alpha<\beta$. Let e_β be the composite

$$r_\beta B \xrightarrow{r_\beta k_\beta} r_\beta s_\beta A_\beta \xrightarrow{\varepsilon_\beta A_\beta} A_\beta \ .$$

from the diagram

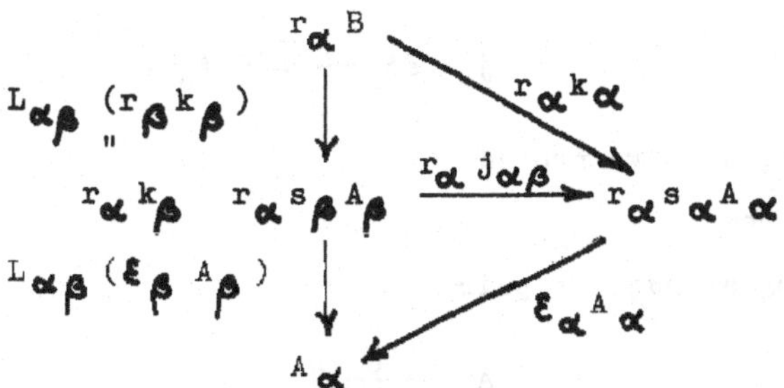

We see that if $\alpha<\beta$ $L_{\alpha\beta}(e_\beta) = e_\alpha$. Since e_α is an isomorphism so is e_β . By transfinite induction, then, we have shown that if A_0 is special it persists.

Let us now assume also that $\underline{B}$ is well-powered. Then from §2 we have the adjoint pair

$$\underleftarrow{\lim}\ \underline{A}_\alpha \overset{r}{\underset{s}{\rightleftarrows}} \underline{B} \quad ,$$

with (ε,η): $r \dashv s$ and ε: $rs \xrightarrow{\sim} 1$. If $A_0 \in \underline{A}_0$ persists, then there is an $A \in \underleftarrow{\lim}\, \underline{A}_\alpha$ such that $L_o A = A_o$. But

$$\varepsilon A\colon rsA \xrightarrow{\sim} A$$

and

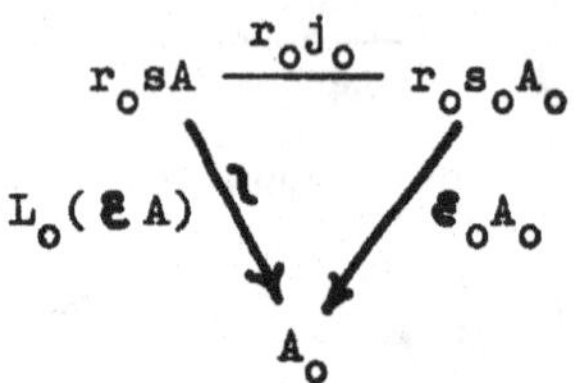

commutes, so A_o is special for the monomorphism

$$j_o\colon sA \longrightarrow a_o A_o \ .$$

Therefore, we have proved

3.3 Theorem.

For an adjoint pair

$$\underline{A}_o \underset{s_o}{\overset{r_o}{\rightleftarrows}} \underline{B} \ .$$

if $\underline{B}$ is complete and well-powered then $A_o \in \underline{A}_o$ is special iff it persists.

The most interesting application of 3.3 is to categories with models [1]. Recall that a category with models is a functor

$$I\colon \underline{M} \longrightarrow \underline{A}_o$$

where $\underline{M}$ is small and $\underline{A}_o$ is cocomplete. I determines an adjoint pair

$$\underline{A}_o \underset{s_o}{\overset{r_o}{\rightleftarrows}} (\underline{M}^*, S)$$

with $(\varepsilon_o, \eta_o): r_o \dashv s_o$ as follows. If $A_o \in \underline{A}_o$ and $M \in \underline{M}$ then

$$s_o A_o(M) = \underline{A}_o(IM, A_o) .$$

If $F \in (\underline{M}^*, S)$, let $\underline{M}_F$ be the category whose objects are pairs, (M,x) where $M \in \underline{M}$ and $x \in FM$. A morphism $\alpha: (M_1,x_1) \to (M_2,x_2)$ is a morphism $\alpha: M_1 \to M_2$ such that $F\alpha(x_2) = x_1$. The projection

$$P_F: \underline{M}_F \longrightarrow \underline{M}$$

is given by $P_F(M,x) = M$, and

$$r_o F = \varinjlim IP_F .$$

If we denote the canonical injection of the colimit by

$$i_x: IP_F(M,x) = IM \longrightarrow r_o F ,$$

then $\varepsilon_o A_o: r_o s_o A_o \to A_o$ is determined by the diagrams

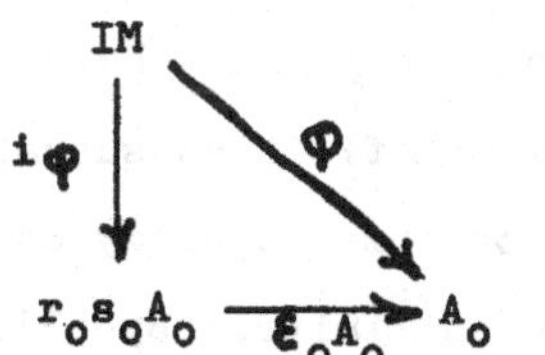

$\eta_o F: F \to s_o r_o F$ is given by

$$\eta_o F(M)(x) = i_x .$$

Now the definition of special for this adjoint pair is precisely the definition of $\underline{M}$-object given in [1], where these are studied at some length. Since $\underline{M}$ is small, $(\underline{M}^*, S)$ is certainly complete and well-powered, so 3.3 tells us that $A_0 \in \underline{A}_0$ is an $\underline{M}$-object iff it persists, and this completely answers the question of the relationship of $\underline{M}$-objects to coalgebras. In [1] this question was evaded by making the assumption that $\varepsilon_1 : r_1 s_1 \longrightarrow 1$ was an equivalence. In this case, of course, the limit is realized at $\underline{A}_1$ and 3.3 degenerates into the statement proved in [1] that $A_0 \in \underline{A}_0$ is an $\underline{M}$-object iff there exists $\theta_0 : A_0 \longrightarrow r_0 s_0 A_0$ such that $(A_0, \theta_0) \in \underline{A}_1$.

From the above we see that the natural definition of the category of $\underline{M}$-objects is $\underline{A} = \lim\limits_{\leftarrow} \underline{A}_\alpha$, or, by the results of §3(i), $(\underline{M}^*, S)[\Sigma_0^{-1}]$ where Σ_0 is the class of natural transformations made invertible by r_0. This is always a coreflective sub-category of $(\underline{M}^*, S)$, making it, for example, complete and cocomplete. This fact indicates clearly the difference between $\underline{M}$-objects and sheaves for a Grothendieck topology on $\underline{M}$ [4]. Namely, by a result of Giraud [4] we know that a coreflective subcategory of $(\underline{M}^*, S)$ is the category of sheaves for some topology on $\underline{M}$ iff the coreflection commutes with finite inverse limits. By [2] it follows that this will be so iff Σ_0 has a calculus of right fractions. The difference, then, is that we are studying arbitrary coreflective subcategories of $(\underline{M}^*, S)$, whereas only those for which the class of transformations made invertible by the coreflection has a calculus of right fractions are considered in [4].

(iii) Completions.

Before starting on our main subject, we need several remarks and definitions concerning model induced adjoint pairs. First, if $\underline{A}$ is cocomplete and $I: \underline{M} \longrightarrow \underline{A}$, then in the model induced adjoint pair

$$\underline{A} \underset{s}{\overset{r}{\leftrightarrows}} (\underline{M}^*, S)$$

defined in (ii), r preserves colimits and satisfies $ry \approx I$, where

$$y: \underline{M} \longrightarrow (\underline{M}^*, S)$$

is the Yoneda embedding. Furthermore, as is immediately evident, these properties determine r up to natural equivalence. Also, since r determines s up to equivalence, the whole adjoint pair is determined by these properties. That is, given any pair

$$\underline{A} \underset{s'}{\overset{r'}{\leftrightarrows}} (\underline{M}^*, S)$$

with $r' \dashv s'$, then $r'y \approx I$ implies that, up to natural equivalence, we are dealing with the adjoint pair induced by I. For the next remark assume that instead of being cocomplete $\underline{A}$ is complete. Then by a technique dual to that of (ii), $I: \underline{M} \longrightarrow \underline{A}$ determines an adjoint pair

$$\underline{A} \underset{{}^*s}{\overset{{}^*r}{\rightleftarrows}} (\underline{M}, S)^*$$

with ${}^*s \dashv {}^*r$. In fact,

$${}^*sA(M) = \underline{A}(A, IM)$$

for $A \in \underline{A}$ and $M \in \underline{M}$ with the obvious effect on morphisms. If $F \in (\underline{M}, S)^*$, let $\underline{M}^F$ be the category whose objects are pairs (M,x) for $M \in \underline{M}$ and $x \in FM$. A morphism $\alpha: (M_1, x_1) \longrightarrow (M_2, x_2)$ is a morphism $\alpha: M_1 \longrightarrow M_2$ in $\underline{M}$ such that $F\alpha(x_1) = x_2$ (F is now covariant). Then we have the evident projection

$$P^F: \underline{M}^F \longrightarrow \underline{M},$$

and

$${}^*rF = \varprojlim IP^F .$$

The obvious dual definitions give the unit and counit ${}^*\eta: 1 \longrightarrow {}^*r\,{}^*s$ and ${}^*\varepsilon: {}^*s\,{}^*r \longrightarrow 1$. As above, ${}^*r$, and hence the whole adjoint pair, is determined by the property of preserving limits and satisfying ${}^*r\,{}^*y \approx I$ where

$${}^*y: \underline{M} \longrightarrow (\underline{M}, S)^*$$

is the contravariant Yoneda embedding.

Suppose now that $\underline{A}$ is, say, complete and cocomplete. Then $I: \underline{M} \longrightarrow \underline{A}$ is called <u>dense</u> if $\varepsilon: rs \longrightarrow 1$ is an equivalence, and <u>codense</u> if ${}^*\eta: 1 \longrightarrow {}^*r\,{}^*s$ is. (This terminology should probably be reversed for obvious reasons, but we use it since it seems to be standard). One says that I <u>generates</u> $\underline{A}$ if ε is epic and <u>cogenerates</u> $\underline{A}$ if ${}^*\eta$ is monic.

With these definitions and remarks, let us get to the matter at hand and assume that

$$I_o : \underline{M} \longrightarrow \underline{A}_o$$

is full, faithful, and codense. To make sense of the latter assumption we may as well assume also that $\underline{A}_o$ is complete, in which case it follows easily that $\underline{A}_o$ is also cocomplete. Moreover, it follows that I_o is cocontinuous. In fact, let M_i be a diagram in $\underline{M}$ such that $\varinjlim M_i$ exists. Then for every $A \in \underline{A}_o$, since ${}^*sI_o \approx {}^*y$ is just another way of saying I_o is full and faithful, we have the following string of equivalences:

$$\begin{aligned}
&\underline{A}(I_o(\varinjlim M_i),\ A) \approx \underline{A}(I_o(\varinjlim M_i),\ {}^*r^*sA) \\
&\approx (\underline{M},S)^*({}^*sI_o(\varinjlim M_i),\ {}^*sA) \approx (\underline{M},\ s)^*({}^*y(\varinjlim M_i),\ {}^*sA) \\
&\approx (\underline{M},S)^*(\varinjlim {}^*yM_i,\ {}^*sA) \approx \varprojlim\ (\underline{M},S)^*({}^*yM_i,\ {}^*sA) \\
&\approx \varprojlim\ (\underline{M},S)^*({}^*sI_oM_i,\ {}^*sA) \approx \varprojlim\ \underline{A}(I_oM_i,\ {}^*r^*sA) \\
&\approx \varprojlim\ \underline{A}(I_oM_i,\ A) \approx \underline{A}(\varinjlim\ I_oM_i,\ A)
\end{aligned}$$

Thus,

$$\varinjlim I_oM_i \xrightarrow{\ \approx\ } I_o(\varinjlim M_i) \ ,$$

and the reader may easily check that this isomorphism is the canonical one. Of course, the dual remark that full, faithful and dense implies continuous is also valid, and it is this form that we shall use later.

Now, using the results of § 2 let us build the tower

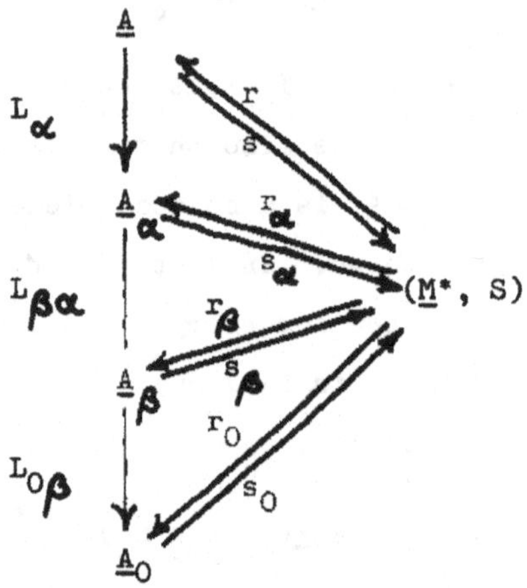

whose 0th component is the model induced pair determined by I_0. Then $r_\alpha \dashv s_\alpha$ for each ordinal α, $\underline{A} = \varprojlim \underline{A}_\alpha$, $r \dashv s$, and $\varepsilon: rs \xrightarrow{\sim} 1$.

For each ordinal $\alpha > 0$ define $I_\alpha : \underline{M} \to \underline{A}_\alpha$ by $I_\alpha = r_\alpha y$, and define $I: \underline{M} \to \underline{A}$ by $I = ry$. Then if $\alpha > \beta > o$ we have $L_{\beta\alpha} I_\alpha = I_\beta$ and $L_\alpha I = I_\alpha$.

By the previous remarks, each adjoint pair $r_\alpha \dashv s_\alpha$ is model induced from I_α as is $r \dashv s$ from I. Also, we have

$L_{o\alpha} I_\alpha = r_o y \approx I_o$, and $L_o I = r_o y \approx I_o$, which implies that I and each I_α is faithful. We remark that there is a way to lift I_o on the nose to each $\underline{A}_\alpha$ and to $\underline{A}$ - for successors see [1] and give the obvious definition for a limit, or use the result of (ii) that "special" implies "persists" after checking the naturality of the constructions with respect to morphisms of $\underline{M}$. However, we do not need such accuracy here.

Now suppose $f: I_\alpha M \to I_\alpha M'$ is a morphism in $\underline{A}_\alpha$. By the above, since I_o is full, there is a $\varphi: M \to M'$ such that $L_{o\alpha}(f) = L_{o\alpha}(I_\alpha \varphi)$. $L_{o\alpha}$ is faithful, so $f = I_\alpha \varphi$ and each I_α is full. The same argument applies, of course, to I. Thus, so far we know that $\underline{A}$ is complete and cocomplete, and I is full, faithful, and dense making I also continuous as above.

Recall from §1 that for $R_{\alpha o} = r_\alpha s_o$ we have $L_{o\alpha} \dashv R_{\alpha o}$, and for $R_o = rs_o$ we have $L_o \dashv R_o$. Thus $I_\alpha \approx R_{\alpha o} I_o$ and $I \approx R_o I_o$ follows from $s_o I_o \approx y$. Now consider the functor

$$ {}^*s_\alpha : \underline{A}_\alpha \longrightarrow (\underline{M}, S)^* \quad . $$

If $A_\alpha \in \underline{A}_\alpha$ and $M \in \underline{M}$ then

$$ {}^*s_\alpha A_\alpha(M) = \underline{A}_\alpha(A_\alpha, I_\alpha M) \approx \underline{A}_\alpha(A_\alpha, R_{\alpha o} I_o M) $$

$$ \approx \underline{A}_o(L_{o\alpha} A_\alpha, I_o M) = \underline{A}_o(A_o, I_o M) = {}^*s_o L_{o\alpha} A_\alpha(M). $$

Since the correspondences are clearly natural in A_α and M, we have

$*s_\alpha \cong *s_o L_{o\alpha}$, and in the same way $*s \cong *s_o L_o$. Therefore, since $L_{o\alpha} \dashv R_{o\alpha}$, $L_o \dashv R_o$, and $*s_o \dashv *r_o$, we have

$$*s_o L_{o\alpha} \dashv R_{\alpha o} r_o$$

and

$$*s_o L_o \dashv R_o {*r_o} \ .$$

By the above, then, we may take $*r_\alpha = R_{\alpha o}{*r_o}$ and $*r = R_o{*r_o}$. From this we see that if $A_\alpha = \underline{A}_\alpha$ then

$$\begin{aligned} R_{\alpha o}L_{o\alpha}A &= R_{\alpha o}A_o \cong R_{\alpha o}({*r_o}{*s_o}A_o) \\ &= {*r_\alpha}{*s_o}A_o = {*r_\alpha}{*s_o}L_{o\alpha}A_\alpha \\ &\cong {*r_\alpha}{*s_\alpha}A_\alpha \ . \end{aligned}$$

It is easy to see that these isomorphisms give an equivalence of triples, so the triple induced on $\underline{A}_\alpha$ by $L_{o\alpha} \dashv R_{o\alpha}$ is equivalent to the model induced triple defined by I_α. The same calculation gives the corresponding result for I. But then, because both L_o and $L_{o\alpha}$ are faithful, I cogenerates $\underline{A}$ and I_α cogenerates $\underline{A}_\alpha$ for each α.

Finally, let M_i be a diagram in $\underline{M}$ for which $\varinjlim M_i$ exists and denote the canonical morphism by

$$c_I\colon \varinjlim IM_i \longrightarrow I(\varinjlim M_i) \ .$$

(In general, for a functor F we write c_F for this comparison morphism). Applying L_o gives a diagram

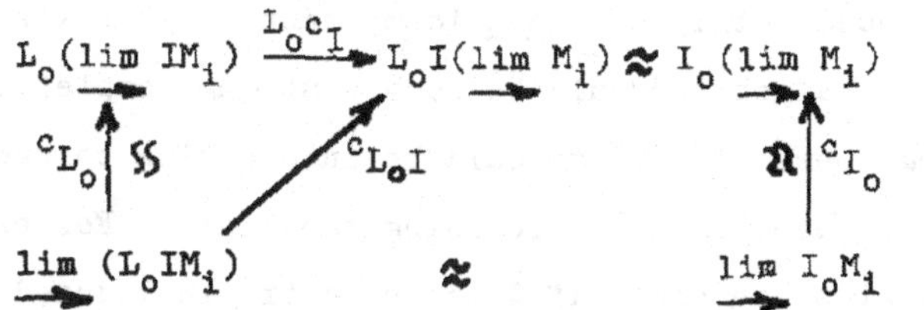

Since L_o reflects isomorphism, c_I is an isomorphism and I is continuous. The same argument applies also to any I_α.

Summing up, we have shown that if I_o is full, faithful, and codense, then I is full, faithful, dense, and cogenerates $\underline{A}$. In addition, $\underline{A}$ is complete and cocomplete and I is continuous and co-continuous. In particular, we may take for I_o

$$*y: \underline{M} \longrightarrow (\underline{M}, S)^*,$$

obtaining in the limit a Lambek completion of $\underline{M}$ [3]. Applying the dual technique to

$$y: \underline{M} \longrightarrow (\underline{M}^*, S)$$

gives a completion with dual properties - i.e., the embedding is codense and generates plus the other self dual properties. The two completions are connected by an adjoint pair whose deviation from an equivalence measures the inability to find a completion for which the embedding is dense <u>and</u> codense - that this is impossible in general is known from examples of Isbell. Thus, *y provides an interesting example of a functor which is full and faithful and whose model induced cotriple is <u>not</u> idempotent - if it were, then the category of coalgebras would be such a strong completion for every $\underline{M}$. Along these lines, it would be interesting to know if the completion is achieved prior to reaching the limit. For example, is it achieved at stage one? If $\mathbb{G}$ is the cotriple induced by

$$^{*}y: \underline{M} \longrightarrow (\underline{M}, S)^{*} ,$$

then

$$^{*}y_1: \underline{M} \longrightarrow (\underline{M}, S)^{*}_{\mathbb{G}}$$

has all the properties of the completion except that it may, possibly, be non-dense. We have considered several examples, among them the counterexamples of Isbell, and for these $^{*}y_1$ does seem to be dense. We did not, however, see a proof of this in the general case.

References

[1] H. Applegate and M. Tierney, Categories with models, Springer Lecture Notes in Mathematics Vol. 80 (1969), pp. 156-244.

[2] P. Gabriel and M. Zisman, Calculus of fractions and homotopy theory, Springer-Verlag, Berlin Heidelberg New York, 1967.

[3] J. Lambek, Completions of categories, Springer Lecture Notes in Mathematics Vol. 24 (1966).

[4] J. L. Verdier, Séminaire de Géometrie Algébrique de l'Institut des Hautes Etudes Scientifiques 1963/64, Fasicule I.

THE RIGHT ADJOINTS INTO THE CATEGORIES OF RELATIONAL SYSTEMS

A. Pultr*

Received June 25, 1969

In [2], Freyd proved that the right adjoints from a cocomplete category $\mathcal{B}$ into a primitive class of algebras $\mathcal{A}$ may be constructed in a natural way from coalgebras over $\mathcal{B}$ satisfying equations dual to the ones defining $\mathcal{A}$. The aim of the present paper is to show that this may be extended to theories more general than the algebraic ones (relational theories - see 2.1). Among the consequences, we show that the right adjoints from a cocomplete $\mathcal{B}$ into the category of relational systems of a given type may be constructed from co-relational systems over $\mathcal{B}$, the right adjoints into the category of transitive relations may be constructed from cotransitive co-relations etc. (see §3).

In §1 the universal $\mathcal{A}\mathcal{B}$--adjunction is defined (it is, roughly speaking, a pair of functors which covers all possible adjoint situations between $\mathcal{A}$ and $\mathcal{B}$) and some results from elsewhere reformulated into a form handy for considerations of §2. In §2 the relational theory is defined and the main theorem (2.5) proved. §3 contains examples of applications of 2.5 (for categories of general relational systems, symmetric relations, transitive relations, partial algebras and Fréchet convergences).

---

*Support from the Canadian National Research Council and the McMaster University is gratefully acknowledged.

<u>Notation</u>: The category of sets and all mappings is denoted by Set, the category of small categories by Cat. The set of all morphisms from a into b in a category $\mathcal{K}$ is denoted by $\langle a,b \rangle_{\mathcal{K}}$ or simply by $\langle a,b \rangle$ (in Set, we use synonymously $\langle X,Y \rangle$ and Y^X). If a is a fixed object of $\mathcal{K}$, $\langle a,- \rangle$ is the functor $\mathcal{K} \to$ Set defined by $\langle a,- \rangle$ (b) = $\langle a,b \rangle$, $(\langle a,- \rangle(\alpha))(\varphi) = \alpha \cdot \varphi$. If F is a functor into $\mathcal{K}$, we write $\langle a,F- \rangle$ for $\langle a,- \rangle \circ F$, if $\varphi: a \to b$ is a morphism, $\langle \varphi,- \rangle$ denotes the natural transformation $\langle b,- \rangle \to \langle a,- \rangle$ etc. The class of objects of $\mathcal{K}$ is denoted by $|\mathcal{K}|$. If $\mathcal{K},\mathcal{C}$ are categories and $a \in |\mathcal{C}|$, $\bar{a}$ (or, more precisely, $\bar{a}^{\mathcal{K}}$) designates the constant functor $\mathcal{K} \to \mathcal{C}$ defined by $\bar{a}(\mu) = 1_a$. Similarly, if $\varphi: a \to b$ is a morphism, $\bar{\varphi}: \bar{a} \to \bar{b}$ is obvious natural transformation. The dual category to $\mathcal{K}$ is denoted by $\mathcal{K}^*$. The ordinal α is always taken as the set of all ordinals less then α(e.g. 2 = $\{0,1\}$, $\omega+1 = \{0,1,2,\dots,\omega\}$ etc.)

§1. <u>Preliminaries</u>

1.1 <u>Definition</u>: Let $\mathcal{A}, \mathcal{B}$ be categories. A <u>universal $\mathcal{A}\mathcal{B}$-adjunction</u> consists of a category $\mathcal{C}$ and functors

$$\underline{L}: \mathcal{A} \times \mathcal{C} \to \mathcal{B}, \quad \underline{R}: \mathcal{C}^* \times \mathcal{B} \to \mathcal{A}$$

such that

1) $\langle \underline{L}(-,-), - \rangle$ and $\langle -, \underline{R}(-,-) \rangle$ are naturally equivalent,

2) if for some category $\mathcal{K}$ and functors L: $\mathcal{A} \times \mathcal{K} \to \mathcal{B}$,

$R: \mathcal{K}^* \times \mathcal{B} \to \mathcal{A}$ there is a natural equivalence $\langle L(-,-), -\rangle \approx \approx \langle -, R(-,-)\rangle$, then there is a functor $H: \mathcal{K} \to \mathcal{C}$ with $L \approx \underline{L}(1 \times H)$, $R \approx \underline{R}(H \times 1)$. (Here the second isomorphism may be chosen as the inverse of the conjugate of the first, with "conjugate" in the sense of Mitchell [2], Prop. 2.1.)

1.2 Remark: Thus, if $(\mathcal{C}, \underline{L}, \underline{R})$ is a universal $\mathcal{A}\mathcal{B}$ -adjunction and if $\mathcal{A} \underset{R}{\overset{L}{\rightleftarrows}} \mathcal{B}$ is a pair of adjoint functors, there is a e $\in |\mathcal{C}|$ such that $L \approx \underline{L}(-,c)$ and $R \approx \underline{R}(c,-)$.

1.3 If A is a small category, denote by $J: A^* \to Set^A$ the Yoneda embedding defined by $J(a) = \langle a,-\rangle$ for objects $a \in |A|$, $J(\alpha)(\varphi) = \varphi \circ \alpha$ for morphisms $\alpha \in A$. Define the functor $K: Set^A \to Cat$ as follows:

for functors $f: A \to Set$ define the category $K(f)$ by

$|K(f)| = \{(a,x) | a \in A,\ x \in f(a)\}$,

$\langle (a,x), (a',x')\rangle_{K(f)} = \{((a,x), \alpha, (a',x')) | \alpha : a' \to a,$

$f(\alpha)(x') = x\}$ (the composition in $K(f)$ is obvious);

For natural transformations $\theta: f \to g$ define the functor $K(\theta)$ by

$$K(\theta)(a,x) = (a, \theta^a(x))$$

$$K(\theta)((a,x), \alpha, (a',x')) = ((a, \theta^a(x)), \alpha (a', \theta^{a'}(x'))).$$

(Thus $K(f)$ is the comma category (p,f), where $p: \mathbb{1} \to Sets$ is a one-point set.)

For $f\colon A \to \mathrm{Set}$ define $\mathcal{U}(f)\colon K(f) \to A^*$ by $\mathcal{U}(f)\,(a,x) = a$, $\mathcal{U}(f)\,((a,x),\alpha,(a',x')) = \alpha$. We have, for any $\Theta\colon f \to f'$,

$$\mathcal{U}(f')\,K(\Theta) = \mathcal{U}(f).$$

Finally, define a natural transformation

$$\tau : J\circ\mathcal{U}(f) \to \bar{f}$$

by $(\tau^{(a,x)})_{a'}(\varphi) = f(\varphi)\,(x)$.
It is a generally known fact that τ is a colimit of the functor $J\circ\mathcal{U}(f)$ (see, e.g., [3], Ch. II, § 1). Thus, J is dense in the sense of Isbell [4] and Ulmer [8].

1.4 Let $\mathcal{B}$ be cocomplete and take $g\colon A^* \to \mathcal{B}$. The commutativity diagram

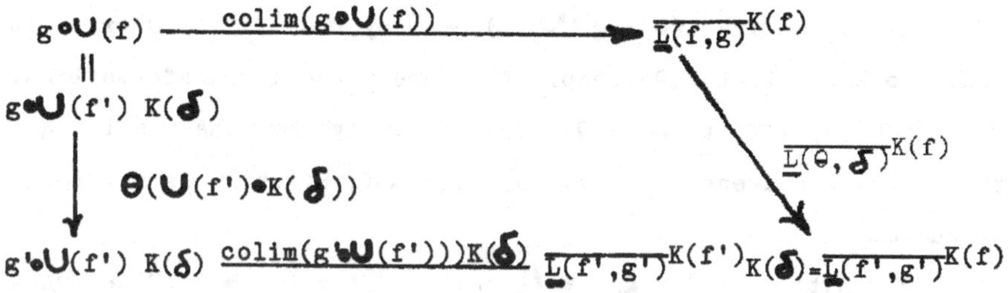

$$\begin{array}{ccc}
g\circ\mathcal{U}(f) & \xrightarrow{\ \mathrm{colim}(g\circ\mathcal{U}(f))\ } & \overline{\underline{L}(f,g)}^{K(f)} \\
\| & & \\
g\circ\mathcal{U}(f')\,K(\delta) & & \Big\downarrow \overline{\underline{L}(\Theta,\delta)}^{K(f)} \\
\Big\downarrow \Theta(\mathcal{U}(f')\circ K(\delta)) & & \\
g'\circ\mathcal{U}(f')\,K(\delta) & \xrightarrow{\ \mathrm{colim}(g'\circ\mathcal{U}(f'))K(\delta)\ } & \overline{\underline{L}(f',g')}^{K(f')}K(\delta) = \overline{\underline{L}(f',g')}^{K(f)}
\end{array}$$

($\delta\colon f \to f'$, $\Theta\colon g \to g'$ are natural transformations) defines, up to a natural equivalence, a functor $\underline{L}\colon \mathrm{Set}^A \times \mathcal{B}^{A^*} \to \mathcal{B}$.

<u>Remark</u>: $\underline{L}(-,g)$ is the Kan extension of g (with respect to J) - see [5], [8].

1.5 Define $\underline{R}\colon (\mathcal{B}^{A^*})^* \times \mathcal{B} \to \mathrm{Set}^A$ putting, for $g\colon A^* \to \mathcal{B}$ and $b \in |\mathcal{B}|$,

$$\underline{R}(g,b) = \langle g\,-,b\rangle,$$

for Θ: $g' \to g$ and φ: $b \to b'$, $\underline{R}(\Theta,\varphi)^a(\mu) = \varphi\circ\mu\circ\Theta^a$.

1.6 The essence of the following statement is, in fact, proved by André in [1]. We shall give another proof here, since transforming of the results of [1] into the form we need would take the same space.

Theorem: In the notation of 1.4 and 1.5, $(\mathcal{B}^{A^*}, \underline{L}, \underline{R})$ is a universal $Set^A\,\mathcal{B}$-adjunction when $\mathcal{B}$ is cocomplete.

Proof: 1) Let f: $A \to Set$, g: $A^* \to \mathcal{B}$, $b \in |\mathcal{B}|$. A one-to-one correspondence between $\langle \underline{L}(f,g), b\rangle$ and $\langle f, \underline{R}(g,b)\rangle$, is by (1.4) just a one-to-one correspondence between the natural transformations $g\circ \mathsf{U}(f) \to \bar{b}$ and the natural transformations $f \to \underline{R}(g,b)$. For Θ: $g\circ \mathsf{U}(f) \to \bar{b}$ define $h(\Theta)$: $f \to \underline{R}(g,b)$ by $h(\Theta)^a(x) = \Theta^{(a,x)}$, for Θ: $f \to \underline{R}(g,b)$ define $t(\Theta)$: $g\circ\mathsf{U}(f) \to \bar{b}$ by $t(\Theta)^{(a,x)} = \Theta^a(x)$. It is easy to check that $h(\Theta)$ resp. $t(\Theta)$ are natural transformations. Obviously $ht(\Theta) = \Theta$, $th(\Theta) = \Theta$. We see easily that the resulting correspondence between $\langle \underline{L}(f,g), b\rangle$ and $\langle f, \underline{R}(g,b)\rangle$ is natural in f,g, and b.

2) Let L: $Set^A \times \mathcal{K} \to \mathcal{B}$ and R: $\mathcal{K}^* \times \mathcal{B} \to Set^A$ be functors such that $\langle L(-,-),-\rangle$ and $\langle -,R(-,-)\rangle$ are naturally equivalent. Define H: $\mathcal{K} \to \mathcal{B}^{A^*}$ by $H(K) = L(J-, K)$ for $K \in |\mathcal{K}|$, $H(\varphi)^a = L(1_{\underline{J}(a)},\varphi)$ for morphisms φ. Since $L(-,K)$ is a left adjoint, it commutes with colimits and we obtain (see 1.4 and 1.3 for τ)

$$\operatorname{colim}(H(K) \circ \mathbf{U}(f)) = \operatorname{colim} L(\underline{J} \circ \mathbf{U}(f)-, K) = L(\mathcal{C}, K)$$

and consequently $\underline{L}(f,H(K))$ is isomorphic to $L(f,c)$. It is easy to see that this isomorphism is natural in f,c. Finally, we have $\underline{R}(H(K),b)\ (a) = \langle H(K)\ (a),\ b\rangle = \langle L(\langle a,-\rangle ,c),\ b\rangle \approx \langle\langle a,-\rangle,\ R(c,b)\rangle \approx R(c,b)\rangle(a)$ by Yoneda lemma. Thus, $R \approx \underline{R} \circ (R \times 1)$.

2. Universal adjunctions for relational theories

2.1 Definition: A relational theory $\underline{T}$ is a couple $(A, \mathcal{D})$ where A is a small category and $\mathcal{D}$ a class of functors $D\colon B_D \to A$ having limits in A.

Remark. This notion was introduced by P. Gabriel in an unpublished paper.

Remark: Thus an algebraic theory in the sense of Lawvere, see e.g. [6], is a particular case of relational theory: A is there the dual of the category of finite sets with some additional morphisms, $\mathcal{D}$ is the class of functors with discrete finite B_D.

2.2 Definition: Let $\underline{T} = (A, \mathcal{D})$ be a theory, $\mathcal{a}$ a complete category. The category $\mathcal{a}^{\underline{T}}$ is the full subcategory of $\mathcal{a}^A$ generated by those functors preserving all the limits of functors from $\mathcal{D}$. If $D\colon B \to A$ is a functor, denote by D^* the functor $B^* \to A^*$ with the same values as D. Put $\mathcal{D}^* = \{D^* \in \mathcal{D}\}$. Let $\mathcal{a}$ be a cocomplete category. Denote by ${}_{\underline{T}}\mathcal{a}$ the full subcategory of $\mathcal{a}^{A^*}$ generated by the functors preserving the colimits of all $D^* \in \mathcal{D}^*$.

2.3 Theorem (Gabriel): $Set^{\underline{T}}$ is a reflective subcategory of Set^{A}.

Remark: This result of Gabriel was communicated to the author by Jon Beck. (See also J. F. Kennison, On Limit Preserving Functors, Ill. J. Math. 12 (1968), 616-619.) The particular case for locally small $Set^{\underline{T}}$ with a cogenerator is obvious by [7] (Ch. V, 3.2), since the embedding $Set^{T} \subset Set^{A}$ evidently preserves limits. Thus, a sceptical reader may add the assumption of local smallness and cogenerator to the statement 2.5 below. In the applications of 2.5 in § 3 always either the validity of this assumption or directly the reflectivity of $Set^{\underline{T}}$ in question is evident.

2.4 Proposition: Let $(\mathcal{C}, \underline{L}, \underline{R})$ be a universal $\mathcal{A}\mathcal{B}$ -adjunction, let $\mathcal{A}_0$ be a reflective subcategory of $\mathcal{A}$, let $J\colon \mathcal{A}_0 \to \mathcal{A}$ be the inclusion, $P\colon \mathcal{A} \to \mathcal{A}_0$ the reflection. Denote by $\mathcal{C}_0$ the full subcategory of $\mathcal{C}$ generated by the $c \in |\mathcal{C}|$ such that $\underline{R}(b,c) \in \mathcal{A}_0$ for every $b \in |\mathcal{B}|$, by C the inclusion $\mathcal{C}_0 \subset \mathcal{C}$. Define

$$\underline{L}_0\colon \mathcal{A}_0 \times \mathcal{C}_0 \to \mathcal{B}, \quad \underline{R}_0\colon \mathcal{C}_0^* \times \mathcal{B} \to \mathcal{A}_0$$

by $\underline{L}_0 = \underline{L} \circ (\underline{J} \times C)$, $J \circ \underline{R}_0 = \underline{R}_0(C \times 1)$.

Then $(\mathcal{C}_0, \underline{L}_0, \underline{R}_0)$ is a universal $\mathcal{A}_0\mathcal{B}$-adjunction.

Proof: Evidently $\langle \underline{L}_0(-,-), - \rangle \approx \langle -, \underline{R}_0(-,-) \rangle$. Let $\langle L(-,-), - \rangle \approx \langle -, R(-,-) \rangle$ for some $L\colon \mathcal{A}_0 \times \mathcal{K} \to \mathcal{B}$, $R\colon \mathcal{K}^* \times \mathcal{B} \to \mathcal{A}_0$. Then also $\langle L(P-,-), - \rangle \approx \langle -, \underline{J} \circ R(-,-) \rangle$, so that there is an $H\colon \mathcal{K} \to \mathcal{C}$ with $L \circ (P \times 1) \approx \underline{L}(1 \times H)$ and $\underline{J} \circ R \approx \underline{R}(H \times 1)$. By the last equivalence, $H = C \circ H_0$ for some $H_0\colon \mathcal{K} \to \mathcal{C}_0$ and we obtain

$$L = L \circ (P \times 1) \circ (J \times 1) \approx \underline{L} \circ (1 \times C \circ H_0) \circ (J \times 1) =$$

$$= L \circ (J \times C) \circ (1 \times H_0) = \underline{L}_0 \circ (1 \times H_0),$$

$$J \circ R \approx \underline{R} \circ (CH_0 \times 1) = J \circ \underline{R}_0 \circ (H_0 \times 1).$$

2.5 __Theorem__: Let $\underline{T} = (A, \mathcal{D})$ be a relational theory, let $\mathcal{B}$ be cocomplete. Then there exist $\underline{L}_0$, $\underline{R}_0$ such that $({}_{\underline{T}}\mathcal{B}, \underline{L}_0, \underline{R}_0)$ is a universal $\mathrm{Set}^{\underline{T}}\mathcal{B}$-adjunction. We may take the restrictions of $\underline{L}$, $\underline{R}$ defined in 1.4, 1.5 for $\underline{L}_0$, $\underline{R}_0$.

__Proof__: By 2.4 and 1.6 it suffices to prove that the objects of ${}_{\underline{T}}\mathcal{B}$ are exactly those objects g of $\mathcal{B}^{A^*}$ for which $\underline{R}(g,b)$ is always in $\mathrm{Set}^{\underline{T}}$. Let g preserve the colimits of D* with $D \in \mathcal{D}$. Using (1.5), we obtain, for any $b \in |\mathcal{B}|$ and $D \in \mathcal{D}$

$$\underline{R}(g,b) \lim D = \langle g \operatorname{colim} D^*, b\rangle = \langle \operatorname{colim} gD^*, b\rangle =$$

$$= \lim \langle gD^*-, b\rangle = \lim (\underline{R}(g,b) \circ D).$$

On the other hand, if $\underline{R}(g,b) = \langle g-, b\rangle$ preserves for every b the limits of all $D \in \mathcal{D}$, we obtain easily

$$\langle g \operatorname{colim} D^*, b\rangle = \langle \operatorname{colim} gD^*, b\rangle$$

for every b, and consequently g colim D* = colim gD*.

A referee has informed the author that the results in his Theorem 2.5 above are also obtained in an independent investigation by John Isbell. They appear in Theorem 3.4 of Isbell's forthcoming paper, "General Functorial Semantics", Part I.

3. Right adjoints into categories of relational systems and other consequences

In this paragraph, several applications of 2.5 are given. The constructions of theories $\underline{T}$ suitable to represent the categories in question as $Set^{\underline{T}}$ are straightforward. We describe it, as an illustration, in the first example. In further examples the explicit descriptions of $\underline{T}$ are omitted.

3.1 Let A be a set. An A-nary relation on a set X is a subset $r \subset X^A$ (we say, of course, binary relation instead of 2-nary relation $r \subset X \times X$). If r resp. s is an A-nary relation on X resp. Y, a mapping $f: X \to Y$ is said to be rs-compatible if $f^A(r) \subset s$ (thus, e.g. in the case of binary relations, if $(x,y) \in r$ implies $(f(x), f(y)) \in s$). A type $\underline{A}$ is a mapping of a set I into the universal class. We write $\underline{A} = (A_i)_I$ to indicate that $\underline{A}(i) = A_i$. A relational system of the type $\underline{A} = (A_i)_I$ on a set X is a system $r = (r_i)_I$ where r_i are A_i-nary relations on X. If r,s are relational systems of the type $\underline{A}$ on X,Y, a mapping $f: X \to Y$ is said to be rs-compatible, or, a homomorphism from (X,r) into (Y,s), if it is $r_i s_i$-compatible for every $i \in I$. The category of all sets with relational systems of the type $\underline{A}$ and all their homomorphisms is denoted by $\mathcal{K}(\underline{A})$.

Now, given a type $\underline{A}$, define a theory $\underline{T} = (A, \vartheta)$ as follows: $|A| = \{\sigma\} \cup I \times 2$ where $\sigma \notin I \times 2$, $\langle (i,1), (i,0) \rangle = \{i\}$ and between other objects there are the morphisms necessary for having

$(i,0) = \sigma^{A_i}\mathcal{B}$ consist of the corresponding power diagrams and the pullbacks

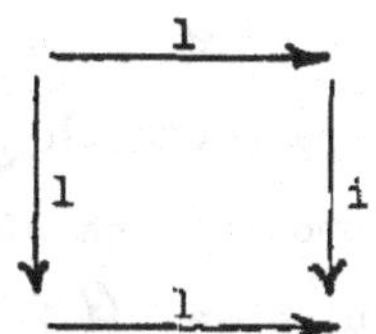

(the latter is here to force the images of i to be monomorphic). Obviously, Set^T is equivalent to $\mathcal{R}(\underline{A})$. Thus, applying 2.5 we obtain the following result, where ${}^{A_i}b$ denotes the coproduct in $\mathcal{B}$ of A_i factors b, with the injections $\nu_a: b \to {}^{A_i}b$.

Proposition: Let $\mathcal{B}$ be cocomplete. Define $\mathcal{C}(\underline{A})$ as follows: $|\mathcal{C}(\underline{A})|$ consists of systems $(b, (\varepsilon_i)_I)$ where $\varepsilon_i: {}^{A_i}b \to e_i$ are epimorphisms in $\mathcal{B}$, the morphisms from $(b, (\varepsilon_i)_I)$ into $(b', (\varepsilon'_i)_I)$ are triples $((\varepsilon_i), \beta, (\varepsilon'_i))$, where $\beta: b \to b'$ is such that there are $\beta_i: e_i \to e'_i$ with $\beta_i \circ \varepsilon_i = \varepsilon'_i \circ {}^{A_i}\beta$ for every i.
Define $\underline{R}_{\underline{A}}: \mathcal{C}(\underline{A}) \times \mathcal{B} \to \mathcal{R}(\underline{A})$ by $\underline{R}_{\underline{A}}((b, (\varepsilon_i)), x) = (\langle b,x\rangle, r)$ where $r = (r_i)_I$, $r_i \subset \langle b,x\rangle^{A_i}$, and

$$\alpha \in r_i \text{ iff there is an } \bar{\alpha}: c_i \to x \text{ with } \alpha(a) = \bar{\alpha} \circ \varepsilon_i \circ \nu_a \text{ for every } a \in A_i.$$

Then there is an $\underline{L}_{\underline{A}}$ such that $(\mathcal{C}(\underline{A}), \underline{L}_{\underline{A}}, \underline{R}_{\underline{A}})$ is a universal $\mathcal{R}(\underline{A})\mathcal{B}$-adjunction.

3.2 Symmetric relations: Let $\mathcal{A}$ be the category of sets with symmetric M-nary relations. Considering the inclusions $\varrho: r \subset X^M$ instead of the relations r (this will be done further on without explicit mention), we see that $(X, \rho) \in |\mathcal{a}|$ are characterized among all objects of $\mathcal{R}((M))$ by

(1) for every permutation $\alpha: M \to M$ there is an $\alpha': r \to r$ such that, for every $m \in M$, $\pi_{\alpha(m)} \circ \rho = \pi_m \circ \rho \circ \alpha'$. (Here π are the projections of the power X^M.

Applying 2.5 we obtain a universal $\mathcal{a}\mathcal{B}$-adjunction $(\mathcal{C}, \underline{L}, \underline{R})$ taking for $\mathcal{C}$ the full subcategory of $\mathcal{C}((M))$ generated by those $(b, \varepsilon: {}^M b \to e)$ with the following property:

(1*) for every permutation $\alpha: M \to M$ there is an $\alpha': e \to e$ such that, for every $m \in M$, $\varepsilon \circ \nu_{\alpha(m)} = \alpha' \circ \varepsilon \circ \nu_m$ and taking for $\underline{L}, \underline{R}$ the restrictions of $L_{(M)}, \underline{R}_{(M)}$.

3.3 <u>Binary transitive relations</u>: These are characterized among the objects of $\mathcal{R}((2))$ by the following:

(2) For the pullback

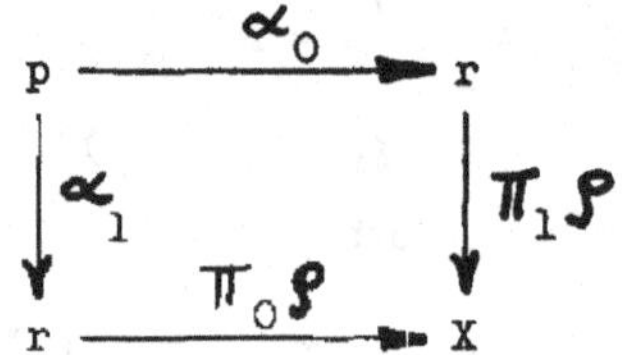

$$\begin{array}{ccc} p & \xrightarrow{\alpha_0} & r \\ \downarrow{\scriptstyle \alpha_1} & & \downarrow{\scriptstyle \pi_1 \rho} \\ r & \xrightarrow{\pi_0 \rho} & X \end{array}$$

there is an $\alpha: p \to r$ with $\pi_i \rho \alpha = \pi_i \rho \alpha_i$ $(i = 0,1)$.

Thus, by 2.5, we obtain a universal $\mathcal{a}\mathcal{B}$-adjunction ($\mathcal{a}$ is now the category of binary transitive relations) $(\mathcal{C}, \underline{L}, \underline{R})$ taking for $\mathcal{C}$ the full subcategory of $\mathcal{C}((2))$ generated by those (b, ε) such that

(2*) For the pushout

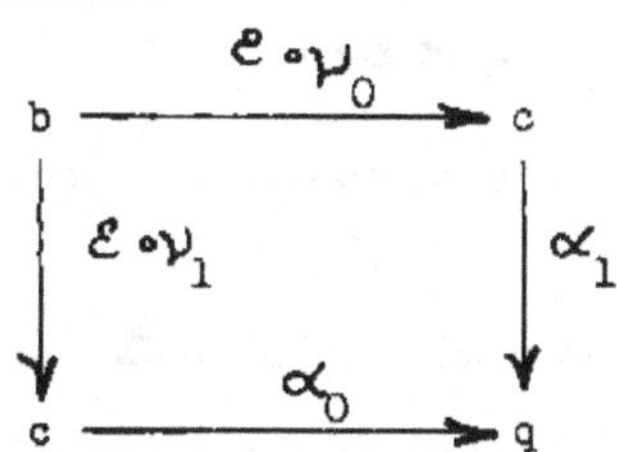

there is an α: $c \to q$ with $\alpha \rho \nu_i = \alpha_i \rho \nu_i$ $(i = 0,1)$.

3.4 <u>Quasialgebras (partial algebras) of a type $\underline{A}$</u>:

For $\underline{A} = (A_i)_I$ define $\underline{A}' = (A'_i)_I$ where $A'_i = A_i \cup \{a_i\}$, $a_i \notin A_i$. The category $Q(\underline{A})$ of <u>quasialgebras</u> of the type $\underline{A}$ is the full subcategory of $\mathcal{R}(\underline{A}')$ generated by the objects $(X, (\rho_i)_I)$ with $\rho_i: r_i \to X^{A_i}$ such that

(3) For $\delta_i: X^{A'_i} \to X^{A_i}$ defined by $\pi_a \circ \delta_i = \pi_a$ for all $a \in A_i$, the $\delta_i \circ \rho_i$ are monomorphisms.

Thus, we obtain a universal $Q(\underline{A})\mathcal{B}$-adjunction $(\mathcal{C}, \underline{L}, \underline{R})$ by taking for $\mathcal{C}$ the full subcategory of $\mathcal{C}(\underline{A}')$ generated by the $(b, (\varepsilon_i)_I)$ such that

(3*) For $\delta'_i: {}^{A_i}b \to {}^{A'_i}b$ defined by $\delta' \circ \nu_a = \nu_a$ for all $a \in A_i$, the $\varepsilon_i\, \delta'_i$ are epimorphisms.

($\mathcal{C}$ is the category of "co-quasialgebras of the type $\underline{A}$ over $\mathcal{B}$").

3.5 <u>Fréchet's convergence</u>: Let ω be the least infinite ordinal, let W be the set of all increasing mappings $\alpha: \omega + 1 \to \omega + 1$. The category $\mathcal{F}$ of Fréchet spaces is the full subcategory of $\mathcal{R}((\omega + 1))$ generated by $(X, \rho : r \to X^{\omega+1})$ such that

(4a) for every $\alpha \in W$ there is an $\alpha' : r \to r$ such that

$$\pi_{\alpha(n)}\rho = \pi_n \rho \alpha' \quad (n \leq \omega),$$

(i.e., the subsequences of convergent sequences converge to the same limits),

(4b) there is a $\beta : X \to r$ with $\rho\beta = \Delta$, where Δ is the diagonal $X \to X^{\omega+1}$. (i.e., each constant sequence x,x,x,... converges to x).

Thus, we obtain a universal $\mathcal{FB}$-adjunction ($\mathcal{C}$, $\underline{L}$, $\underline{R}$) taking for $\mathcal{C}$ the full subcategory of $\mathcal{C}((\omega + 1))$ generated by the objects $(b, \mathcal{E} : {}^{(\omega+1)}b \to c)$ such that

(4a*) for every $\alpha \in W$ there is an $\alpha' : c \to c$ such that

$$\mathcal{E}\nu_{\alpha(i)} = \alpha' \circ \mathcal{E} \circ \nu_n ,$$

(4b*) there is a $\beta : c \to b$ with $\beta\mathcal{E} = \nabla$.

(For the case of the category of Fréchet spaces with unique limits combine 3.5 and 3.4.)

References

[1] M. André: Categories of functors and adjoint functors, Am. J. of Math., 88 (1966), 529-543.

[2] P. Freyd: Algebra valued functors in general and tensor products in particular, Colloquium Mathematicum XIV (1966), 89-106.

[3] P. Gabriel, M. Zisman: Calculus of Fractions and Homotopy Theory, Springer Verlag, New York, 1967.

[4] J. Isbell: Structure of categories, Bull. AMS 72 (1966), 619-655.

[5] D. M. Kan: Adjoint functors, Trans. AMS 87 (1958), 295-329.

[6] F. E. J. Linton: Some aspects of equational categories, Proc. of the Conference of Categorical Algebra (La Jolla 1965), Springer Verlag, New York 1966.

[7] B. Mitchell: Theory of Categories, Academic Press, New York and London 1965.

[8] F. Ulmer: Properties of dense and relative adjoint functors, J. Algebra 8 (1968), 77-95.

TRIADS IN THE HOMOLOGY OF CATEGORIES

by S. Świerczkowski

Received April 22, 1969,

Revised Sept. 30, 1969

Let $\underline{N}$ be a category, $\underline{M} \subset \underline{N}$ a full and small subcategory and $T: \underline{M} \to \underline{A}$ a functor, where $\underline{A}$ is an abelian category such that every set of objects of $\underline{A}$ has a sum (i.e. coproduct) in $\underline{A}$. Let

$$H_k(- , T): \underline{N} \to \underline{A}; \; k = 0, 1, 2, \ldots$$

be the corresponding homology functors defined by M. André in [1].

Let $\underline{C}$ be a category, and $G = (\epsilon, \mathcal{E}, \delta)$ a cotriad on $\underline{C}$ (we use the shorter and less ambiguous "triad" instead of the more usual "triple"; the term "triad" is a suggestion of Saunders Mac Lane). Let $\underline{C}' \subset \underline{C}$ be a full subcategory such that Gc is an object in $\underline{C}'$ whenever c is an object in $\underline{C}$, and let $\underline{E}: \underline{C}' \to \underline{A}$ be a functor, where $\underline{A}$ is an abelian category. Denote by

$$H_k(- , E)_{\mathbf{G}}: \underline{C} \to \underline{A} \; ; \; k = 0, 1, 2, \ldots$$

the homology functors defined by M. Barr and J. Beck in [2].

The purpose of this note is to show that the homology of André is a special case of the homology of Barr and Beck. That this is so under an additional hypothesis has been shown by F. Ulmer [5]. We shall prove the

THEOREM. For every $\underline{N} \supset \underline{M} \xrightarrow[T]{} \underline{A}$ as above, there exists a category $\underline{C}$, a cotriad $\mathbb{G}$ on $\underline{C}$, a full subcategory $\underline{C}' \subset \underline{C}$ containing all objects Gc, where c is an object of $\underline{C}$, a functor $\Phi: \underline{N} \to \underline{C}$ and functor $E: \underline{C}' \to \underline{A}$, such that

$$H_k(- , T) = H_k(\Phi(-), E)_{\mathbb{G}} \; ; \; k = 0, 1, 2, \ldots$$

as functors $\underline{N} \to \underline{A}$.

The first two §'s below are devoted to a brief recalling of some definitions from [1] and [2].

I am grateful to the referee for suggestions concerning the presentation of my proof.

1. André homology [1]

Given $\underline{N} \supset \underline{M} \xrightarrow[T]{} \underline{A}$ as above, let n be an object in $\underline{N}$, and denote by $I_k(n,T)$; $k = 0,1,2, \ldots$, the set of all $(k+1)$ - tuples $(\beta, \alpha_1, \ldots, \alpha_k)$ such that

$$n \xleftarrow[\beta]{} m_0 \xleftarrow[\alpha_1]{} m_1 \leftarrow \cdots \leftarrow m_{k-1} \xleftarrow[\alpha_k]{} m_k ,$$

where all m_i are in $\underline{M}$. Define further

$$\langle \beta, \alpha_1, \ldots, \alpha_k \rangle = Tm_k \; , \; \langle \beta \rangle = Tm_0 \; ,$$

and denote by $C_k(n,T)$ the coproduct $\sum \langle \beta, \alpha_1, \ldots, \alpha_k \rangle$ with $(\beta, \alpha_1, \ldots, \alpha_k)$ ranging over $I_k(n,T)$. Recall that if for every summand $\langle \beta, \alpha_1, \ldots, \alpha_k \rangle$ of $C_k(n,T)$ there is given a morphism to some summand of $C_{k-1}(n,T)$, then this defines in an obvious way a morphism $C_k(n,T) \to C_{k-1}(n,T)$. Thus let $s_k^i\colon C_k(n,T) \to C_{k-1}(n,T)$ be for $i = 0,\dots,k-1$ the morphism sending $\langle \beta, \alpha_1, \dots, \alpha_k \rangle$ to $\langle \beta, \alpha_1, \dots, \alpha_i\alpha_{i+1}, \dots, \alpha_k \rangle$ by the identity $Tm_k \to Tm_k$ (for $i = 0$, $\alpha_i\alpha_{i+1}$ means $\beta\alpha_1$). Let s_k^k send $\langle \beta, \alpha_1, \dots, \alpha_k \rangle$ to $\langle \beta, \alpha_1, \dots, \alpha_{k-1} \rangle$ by $T\alpha_k$. Setting $d_k = \sum (-1)^i s_k^i$, one obtains a chain complex

$$\dots \to C_2(n,T) \xrightarrow[d_2]{} C_1(n,T) \xrightarrow[d_1]{} C_0(n,T) \to 0$$

whose homology is $H_k(n,T)$.

2. Barr and Beck homology [2]

Let $\mathbb{G} = (G, \varepsilon, \delta)$ be a cotriad on $\underline{C}$. Then for every c in $\underline{C}$ there is a morphism $\varepsilon_c\colon Gc \to c$. Denote by $(\varepsilon_i^{k+1})_c\colon G^{k+1}c \to G^k c$; $i = 0,1,2,\dots,k$, (where G^k is the iterated functor), the morphisms.

$$(\varepsilon_i^{k+1})_c = G^{k-i}\varepsilon_{G^ic}\colon G^{k-i}G(G^ic) \to G^{k-i}(G^ic) .$$

If $\underline{C}' \subset \underline{C}$ is a full subcategory containing all Gc for c in $\underline{C}$, and $E\colon \underline{C}' \to \underline{A}$ is a functor into some abelian category $\underline{A}$, then we have for every c in $\underline{C}$ a chain complex

$$\dots \to EG^3c \xrightarrow[\partial_2]{} EG^2c \xrightarrow[\partial_1]{} EGc \to 0 ,$$

where $\partial_k = \sum (-1)^i E(\varepsilon_i^{k+1})_c$. Its homology is denoted by $H_k(c,E)_G$.

3. Adjoint functors

We define first the category $\underline{C}$, next a pair of adjoint functors $\underline{B} \xrightarrow[F]{} \underline{C} \xrightarrow[U]{} \underline{B}$ from which the required cotriad will be obtained.

We shall use the comma category of a pair (H',H"); recall that such a category is associated to a pair of functors $\underline{K}' \xrightarrow[H']{} \underline{K} \xleftarrow[H'']{} \underline{K}''$ and has for its objects all triples $(k', H'k' \xrightarrow[\beta]{} H''k'', k'')$ where k',k" are objects in $\underline{K}'$,$\underline{K}''$ and β is in $\underline{K}$. Its morphisms are pairs $(k' \xrightarrow[\alpha']{} k_1', k'' \xrightarrow[\alpha'']{} k_1'')$ such that

$$\begin{array}{ccc} H'k' & \xrightarrow[\beta]{} & H''k'' \\ \downarrow H'\alpha' & & \downarrow H''\alpha'' \\ H'k_1' & \xrightarrow[\beta_1]{} & H''k_1'' . \end{array}$$

(All diagrams drawn are assumed to commute). There is a functor Ψ from the comma category to $\underline{K}'$ defined by $\Psi(k',\beta,k'') = k'$, $\Psi(\alpha',\alpha'') = \alpha'$. It will be called the first canonical functor.

Assume the notation of §1. If f is a functor with dom f a small, discrete category and codom f = $\underline{N}$, then the comma category of ($\underline{M} \subset \underline{N}$, f) will be more briefly denoted by $|f|$. Thus an object of $|f|$ is of the form $(m, m \xrightarrow[\beta]{} f(x), x)$ where m,x are objects in $\underline{M}$, dom f respectively, and β is a morphism in $\underline{N}$. Such object of $|f|$ will be briefly denoted by (x,β). Write $\Psi_f : |f| \rightarrow \underline{M}$ for the first canonical functor, so that $\Psi_f(x,\beta) = \text{dom}\,\beta$ on the objects. Since $\underline{M}$ is small, so is $|f|$.

Let $\underline{\text{Kit}}$ be the category of small categories. $\underline{C}$ is now defined to be the subcategory of $\underline{\text{Kit}}$ whose objects are all the categories of the form $|f|$, as above, and such that a morphism in $\underline{\text{Kit}}$ (i.e., a functor) $\Psi : |f| \to |g|$ is a morphism in $\underline{C}$ iff

$$\text{(1)} \qquad \begin{array}{ccc} |f| & \xrightarrow{\Psi} & |g| \\ & \Psi_f \searrow \quad \swarrow \Psi_g & \\ & \underline{M} & \end{array} .$$

Let D: $\underline{\text{Kit}}$ — $\underline{\text{Kit}}$ be the functor which assigns to each category c the discrete category Dc with the same object set as c, and which sends each functor $\Psi : c \to c'$ to the object map $d\Psi$ of Ψ.

PROPOSITION 1. A morphism $\omega : D|f| \to D|g|$ is of the form $\omega = D\Psi$ for some $\Psi : |f| \to |g|$ in $\underline{C}$ iff

$$(*) \qquad \omega(x,\gamma) = (y,\beta) \Rightarrow \omega(x,\gamma\alpha) = (y,\beta\alpha)$$

for every $(x,\gamma) \in |f|$, $(y,\beta) \in |g|$ and α in $\underline{M}$ such that $\gamma\alpha$ exists. Moreover there is then exactly one $\Psi : |f| \to |g|$ in $\underline{C}$ such that $\omega = D\Psi$.

The proof follows easily from (1) and the fact that dom f and dom g are discrete categories. As a corollary, one obtains that a morphism $\Psi : |f| \to |g|$ in $\underline{C}$ is uniquely determined by the values it takes on the objects of $|f|$, and to define Ψ it suffices to specify these values so that (*) holds (for Ψ in place of ω).

A functor f with dom f a small discrete category and codom f = $\underline{N}$ will be called a <u>set over</u> $\underline{N}$. By a morphism $\varkappa$: f $\to$ g of sets over $\underline{N}$ we shall mean a functor $\bar{\varkappa}$ satisfying the diagram

$$(2) \qquad \begin{array}{ccc} \text{dom } f & \xrightarrow{\bar{\varkappa}} & \text{dom } g \\ & {\scriptstyle f}\searrow \quad \swarrow{\scriptstyle g} & \\ & \underline{N} & \end{array} .$$

The category $\underline{Sets}/\underline{N}$ may be also thought of as the comma category of D: $\underline{Kit} \to \underline{Cat} \leftarrow \underline{1}$, where the right functor sends 1 to $\underline{N}$. Define a functor $\tilde{F}$: $\underline{Sets}/\underline{N} \to \underline{C}$ as follows: $\tilde{F}f = |f|$, and for $\varkappa$: f $\to$ g as above let $\tilde{F}\varkappa$: $|f| \to |g|$ satisfy $(\tilde{F}\varkappa)(x,\beta) = (\bar{\varkappa}(x),\beta)$ for each $(x,\beta) \in |f|$. Due to Proposition 1, $\tilde{F}\varkappa$ is a morphism in $\underline{C}$.

A set over $\underline{N}$ which factorizes through $\underline{M} \subset \underline{N}$ will be called a set over $\underline{M}$. Let $\underline{B} \subset \underline{Sets}/\underline{N}$ denote the full subcategory whose objects are the sets over $\underline{M}$. Let F: $\underline{B} \to \underline{C}$ be the restriction of $\tilde{F}$. Then F has a left adjoint U: $\underline{C} \to \underline{B}$ defined as follows. For each $|f|$ in $\underline{C}$,

$$U|f|: D|f| \xrightarrow[D\Psi_f]{} D\underline{M} \subset \underline{M} \qquad \text{(see (1))}.$$

For each Ψ : $|f| \to |g|$ in $\underline{C}$, $U\Psi$ is given by

$$\overline{U\Psi} = D\Psi: D|f| \to D|g| \qquad \text{(see (2))}.$$

For each f in $\underline{B}$, let η_f: f $\rightarrow$ U|f| be the morphism defined by $\bar{\eta}_f(x) = (x,1)$, where $x \in$ dom f and 1 is the identity $f(x) \rightarrow f(x)$. It is easy to check that the η_f constitute a natural transformation η: $I_{\underline{B}} \rightarrow UF$. The following lemma states that each η_f is a "universal arrow from f to U". This is well known [4] to be equivalent to the adjointness of (F,U).

LEMMA 1. For every morphism κ : f $\rightarrow$ U|g| in $\underline{B}$, there is a unique Ψ : Ff $\rightarrow$ |g| in $\underline{C}$ such that

$$\begin{array}{ccc} f & \xrightarrow{\eta_f} & UFf \\ & \searrow^{\kappa} & \downarrow U\Psi \\ & & U|g| \end{array} \quad . \tag{3}$$

Proof. $U\Psi$ was defined by $\overline{U\Psi} = D\Psi$: D|f| $\rightarrow$ D|g|. Thus it will suffice to show the existence and uniqueness of Ψ : |f| $\rightarrow$ |g| in $\underline{C}$ such that

$$\begin{array}{ccc} \text{dom } f & \xrightarrow{\bar{\eta}_f} & D|f| \\ & \searrow^{\bar{\kappa}} & \downarrow D\Psi \\ & & D|g| \end{array} \quad . \tag{4}$$

Denote, for every $x \in$ dom f, $\bar{\kappa}(x)$ by (y_x, β_x), where $y_x \in$ dom g.
Uniqueness. Suppose Ψ exists. By (4), $\Psi(x,1) = (y_x, \beta_x)$. Thus Proposition 1 implies $\Psi(x,\alpha) = \Psi(x,1\cdot\alpha) = (y_x, \beta_x\alpha)$ for every α in $\underline{M}$ with codom α = f(x). This determines Ψ uniquely on all

objects of $|f|$, whence by Proposition 1, Ψ is also determined on the morphisms of $|f|$.

Existence. Define Ψ on the objects of $|f|$ by $\Psi(x,\alpha) = (y_x, \beta_x\alpha)$. By (*), Proposition 1, Ψ is uniquely defined as a morphism in $\underline{C}$. Evidently (4) holds.

The morphisms Ψ and κ appearing in (3) will be said to correspond to each other by adjointness.

4. Proof of the theorem

Let $\mathfrak{G} = (G, \varepsilon, \delta)$ be the cotriad corresponding to (F,U). Then $G = FU$ and $\varepsilon\colon FU \to I_{\underline{C}}$ is the natural transformation such that for every $|f|$ in $\underline{C}$, $\varepsilon_{|f|}$ corresponds by adjointness to $\kappa = 1$, i.e.,

$$(5)\qquad \begin{array}{ccc} U|f| & \xrightarrow{\eta_{U|f|}} & UFU|f| \\ & {\scriptstyle \kappa=1}\searrow & \downarrow{\scriptstyle U\varepsilon_{|f|}} \\ & & U|f| \end{array} \;.$$

PROPOSITION 2. For $|f_0| \in \underline{C}$ denote by $f_k \in \underline{B}$; $k = 1,2,3,\ldots$ the functors such that

(a) the objects of dom f_k are all ξ of the form

$$\xi = ((\ldots((x,\beta),\alpha_1),\ldots,\alpha_{k-2}),\alpha_{k-1}) ,$$

where $x \in \operatorname{dom} f_0$; dom β, $\alpha_1,\ldots,\alpha_{k-1}$ are in $\underline{M}$, and $\beta\alpha_1\alpha_2\ldots\alpha_{k-1}$ exists,

(b) $f_k(\xi) = \mathrm{dom}\,\alpha_{k-1}$ (or dom β if $k = 1$) for the above ξ. Then $f_{k+1} = U|f_k|$ and $|f_k| = G^k|f_0|$ for $k = 0,1,2,\ldots$.

The proof is by induction, starting from the easy case $f_1 = U|f_0|$. Having shown $f_{k+1} = U|f_k|$, one has $|f_{k+1}| = FU|f_k| = G|f_k|$, whence $|f_k| = G^k|f_0|$. Subsequently we shall denote the above ξ by $(x,\beta,\alpha_1,\ldots,\alpha_{k-1})$.

We evaluate $\varepsilon_{|f|}$ on an object $\xi \in G|f|$. By Proposition 2 (with $f_0 = f$), $G|f| = |f_1|$, $U|f_1| = f_2$, whence $DG|f| = D|f_1|$, $D|f_1| = \mathrm{dom}\,f_2$. Thus $\xi = (x,\beta,\alpha_1) \in \mathrm{dom}\,f_2$. From (5) it follows that $K(x,\beta) = (x,\beta)$. The proof of Lemma 1 (Existence) implies now that $\varepsilon_{|f|}(x,\beta,\alpha_1) = (x,\beta\alpha_1)$.

We define presently the functor Φ: $\underline{N} \to \underline{C}$. For each $n \in \underline{N}$ consider the discrete subcategory of $\underline{N}$ composed of the object n and denote by $\bar{n}$ its inclusion functor into $\underline{N}$. Put $\Phi n = |\bar{n}|$. The objects of $|\bar{n}|$ are the (n,γ) where dom $\gamma \in \underline{M}$, codom $\gamma = n$. For β: $n \to n'$, let $\Phi\beta$: $|\bar{n}| \to |\bar{n}'|$ be the functor given on the objects of $|\bar{n}|$ by $(\Phi\beta)(n,\gamma) = (n',\beta\gamma)$ (cf. Proposition 1). It may be worth noting that Φ is full and faithful iff $\underline{M}$ is adequate [3] in $\underline{N}$.

Let $\underline{C}'$ be the full subcategory of $\underline{C}$ whose objects are all the $|f|$, where $f \in \underline{B}$. For the purpose of defining E: $\underline{C}' \to \underline{A}$ note that an $|f| \in \underline{C}'$ determines f uniquely: $x \in \mathrm{dom}\,f$ iff $(x,\alpha) \in |f|$ for some α, and $f(x) = \mathrm{codom}\,\alpha$. Thus define E on the objects of $\underline{C}'$ by

$$(6) \qquad E|f| = \sum_{x \in \mathrm{dom}\, f} Tf(x) .$$

Given $\Psi: |f| \to |g|$ in $\underline{C}'$, take any $x \in$ dom f, consider $(x,1) \in |f|$ and denote $\Psi(x,1)$ by (y,β). Then $\beta: f(x) \to f(y)$, by (1). Now let $E\Psi: E|f| \to E|g|$ be the morphism which sends the summand $Tf(x)$ to $Tg(y)$ by $T\beta$. If $\Psi' : |g| \to |h|$ and $\Psi'(y,1) = (z,\gamma)$, then $\Psi'\Psi(x,1) = \Psi'(y,\beta) = (z,\gamma\beta)$ by (*), Proposition 1. This implies $E(\Psi'\Psi) = E\Psi' E\Psi$. The following lemma concludes our proof.

<u>LEMMA 2</u>. For every $n \in \underline{N}$ and $k = 0,1,2,\ldots$ we have

$$EG^{k+1}\Phi n = C_k(n,T) \text{ and } s_k^i = E(\varepsilon_i^{k+1})_{|\bar{n}|};\ i = 0,1,\ldots,k$$

(cf. notation of §'s 1 & 2).

<u>Proof</u>. Put $f_0 = \bar{n}$ in Proposition 2. Then $|f_{k+1}| = G^{k+1}|\bar{n}| = G^{k+1}\Phi n$. Thus the first part of the assertion becomes an immediate consequence of (6).

We have $(\varepsilon_i^{k+1})_{|\bar{n}|} = G^{k-i}\,\varepsilon_{|f_i|}: |f_{k+1}| \to |f_k|$. Above we evaluated $\varepsilon_{|f|}$ on the objects of $G|f|$. Thus

$$\varepsilon_{|f_i|}(n,\beta,\alpha_1,\ldots,\alpha_{i+1}) = (n,\beta,\alpha_1,\ldots,\alpha_{i-1},\alpha_i\alpha_{i+1})$$

for each $(n,\beta,\ldots,\alpha_{i+1}) \in |f_{i+1}|$. Consequently

$$(7) \quad (G^{k-i}\varepsilon_{|f_i|})(n,\beta,\alpha_1,\ldots,\alpha_{k+1}) = (n,\beta,\alpha_1,\ldots,\alpha_i\alpha_{i+1},\ldots,\alpha_{k+1}).$$

Let $x = (n,\beta,\alpha_1,\ldots,\alpha_k) \in$ dom f_{k+1} so that $Tf_{k+1}(x)$ is a summand of $E|f_{k+1}| = C_k(n,T)$. Denote $(G^{k-i}\varepsilon_{|f_i|})(x,1)$ by (y,β). If $i < k$, then (7) implies $y = (n,\beta,\alpha_1,\ldots,\alpha_i\alpha_{i+1},\ldots,\alpha_k)$ and $\beta = \alpha_{k+1} = 1$. Hence $E(G^{k-i}\varepsilon_{|f_i|})$ sends $Tf_{k+1}(x)$ to $Tf_k(y)$ by the identity morphism, which is the same as does s_k^i. If $i = k$, then (7) implies $y = (n,\beta,\alpha_1,\ldots,\alpha_{k-1})$, $\beta = \alpha_k$. Thus $E\,\varepsilon_{|f_k|}$ sends $Tf_{k+1}(x)$ to $Tf_k(y)$ by $T\alpha_k$, which is again the same as does s_k^k.

References

[1] Michel André, Méthode Simpliciale en Algèbre Homologique et Algèbre Commutative, Lecture Notes, No. 32, Springer 1967.

[2] M. Barr and J. Beck, Homology and standard constructions, Seminar on Triples and Categorical Homology Theory, Lecture Notes No. 80, Springer 1969, 245-335.

[3] J. R. Isbell, Adequate subcategories, Illinois J. Math. 4 (1960), 541-552.

[4] Saunders Mac Lane, Categorical Algebra, Bull. Amer. Math. Soc. 71 (1965), 40-106.

[5] F. Ulmer, On cotriples and André (co)homology, their relationship with classical homological algebra, Seminar on Triples and Categorical Homology Theory, Lecture Notes No. 80, Springer 1969, 376-398.

The Australian National University,
Canberra.

Dinatural Transformations

by

Eduardo Dubuc and Ross Street

Received Nov. 17, 1969

This note will introduce the concept of dinatural transformation. This name was chosen because dinaturals are defined between functors of two variables and the components are arrows between the values of the functors at pairs of objects on the diagonal. Dinaturals provide a common setting for naturals and the two types of extraordinary naturals [2]. Moreover, the concept provides a notion of naturality in situations where previously none had been considered. We hope the examples which appear at the end will convince the reader that dinaturals turn up throughout category theory.

Definition. Suppose A,B are categories and $H,K : A^{op} \times A \longrightarrow B$ are functors. A dinatural transformation $\theta : K \longrightarrow K$ is a family of arrows $\theta_a : H(a,a) \longrightarrow K(a,a)$, $a \in A$, in B such that, for all arrows $f : a \longrightarrow b$ in A, the following hexagon commutes.

(1)

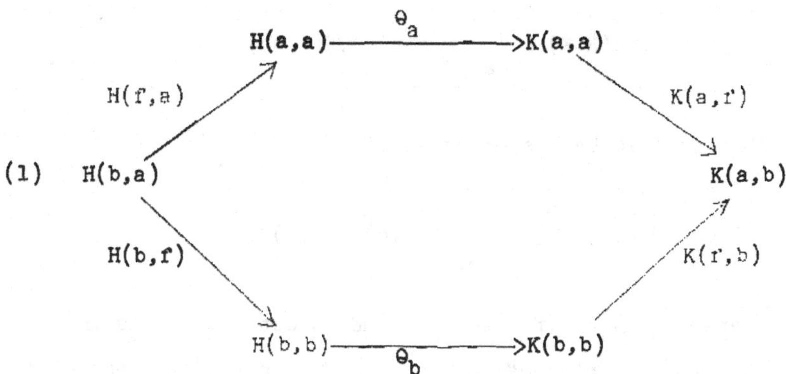

A transformation $\alpha : H \longrightarrow K$ which is natural in both variables, is dinatural; that is, the family $\theta = (\theta_a)$ given by $\theta_a = \alpha_{a,a}$ is dinatural. The (evaluationwise) composite of two dinatural transformations need not be dinatural. If either is a natural transformation (in both variables) then the composite is dinatural.

Theorem 1. Suppose $H,K : A^{op} \times A \longrightarrow B$ are functors where A is small. The elements of the set

$$\int_a B(H(a,a), K(a,a))$$

which is the end of the functor

$$B(\ H(-,-),\ K(-,-))\ :\ A^{op} \times A \longrightarrow \mathcal{S}$$

(where $\mathcal{S}$ is the category of sets) are in one-to-one correspondence with the dinatural transformations from H to K .//

See [1] for the definition of end. Recall, also from [1], that

$$\text{(2)} \qquad Nat(H,K) \cong \int_{a,b} B(H(a,b),K(a,b)).$$

Theorem 1 is expressed by a similar equation

$$Dinat(H,K) \cong \int_{a} B(H(a,a),K(a,a)).$$

The cotensor $x \pitchfork b$ of a set x and object $b \in B$ is the product $\prod_x b$ of x copies of b in B. Cotensor is functorial in x and b. For each element of x there is a projection $x \pitchfork b \longrightarrow b$.

Theorem 2. Suppose $K : A^{op} \times A \longrightarrow B$ is a functor and, for each pair a,b of objects of A, the expression

$$\int_{c} (A(b,c) \times A(c,a)) \pitchfork K(c,c)$$

exists in B (for example, if A is small and B is complete). Then there exists a functor $\overline{K} : A^{op} \times A \longrightarrow B$ (whose value at (a,b) is given by the above expression) and a dinatural transformation $\epsilon : \overline{K} \longrightarrow K$ such that any other dinatural transformation $\Theta : H \longrightarrow K$ factors as

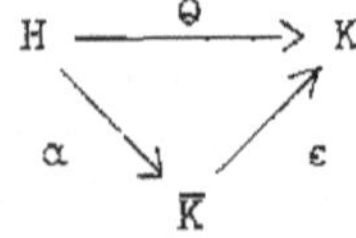

<u>for some unique natural transformation</u> $\alpha : H \longrightarrow \bar{K}$.

<u>Proof.</u> For $(a,b) \in A^{op} \times A$, let

(4) $$\bar{K}(a,b) = \int_c (A(b,c) \times A(c,a)) \pitchfork K(c,c).$$

For arrows $u:c \longrightarrow a, v:b \longrightarrow c$ in A there is a projection from the cotensor under the integral to $K(c,c)$. The composite of this with the canonical projection from the end will be denoted by $p_{u,v} : \bar{K}(a,b) \longrightarrow K(c,c)$. The universal property of these projections is: for each commutative diagram

(5)

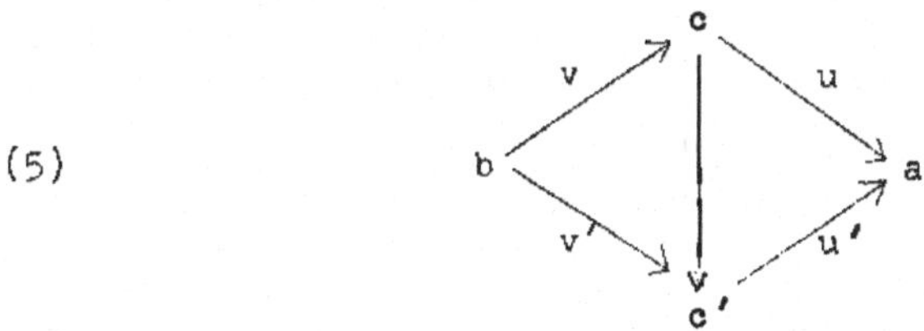

in A, there is a commutative diagram

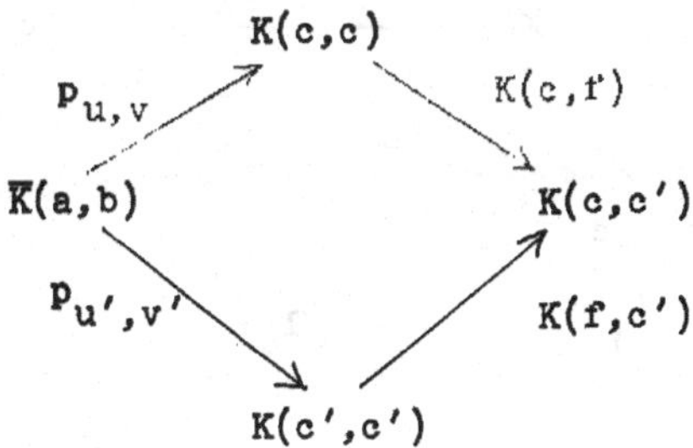

in B. If $m:a \longrightarrow a'$ $n:b' \longrightarrow b$ are arrows of A, then $\bar{K}(m,n) : \bar{K}(a',b') \longrightarrow \bar{K}(a,b)$ is defined by the commutative diagrams

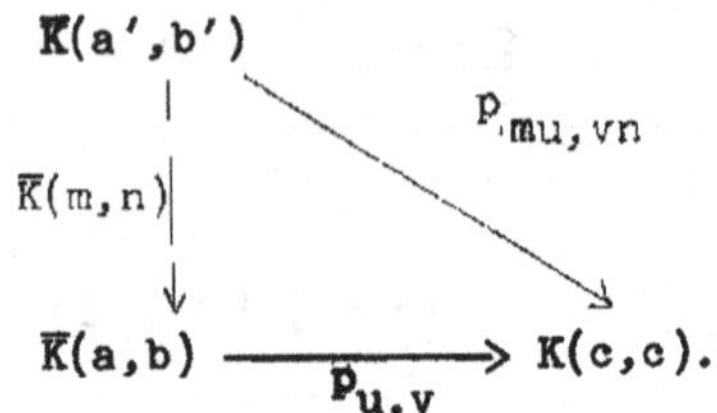

Then $\overline{K}: A^{op} \times A \longrightarrow B$ is a functor. For $c \in A$, let $\epsilon_c = p_{1_c,1_c}$: $\overline{K}(c,c) \longrightarrow K(c,c)$. If $f: c \longrightarrow c'$ is an arrow of A then the diagram

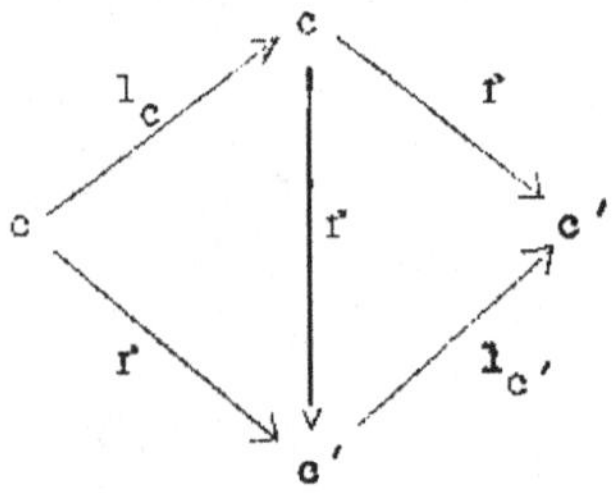

commutes in A; it follows that the diamond in the following diagram commutes.

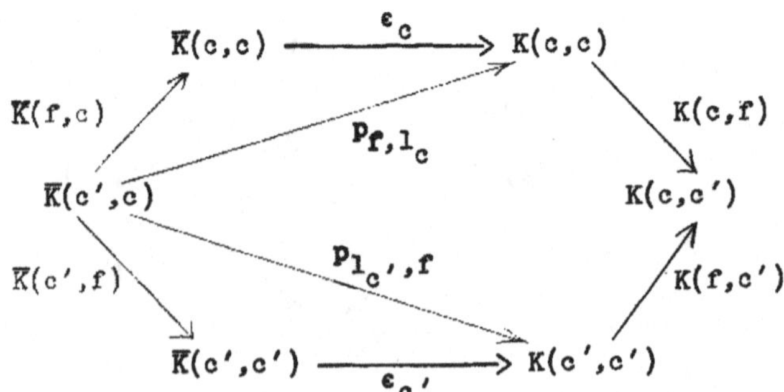

The two triangles commute by definition of $\overline{K}$ on arrows. So $\epsilon = (\epsilon_c) : \overline{K} \longrightarrow K$ is a dinatural transformation.

Suppose $\Theta : H \longrightarrow K$ is dinatural. Whenever (5) commutes in A, the following diagram commutes in B.

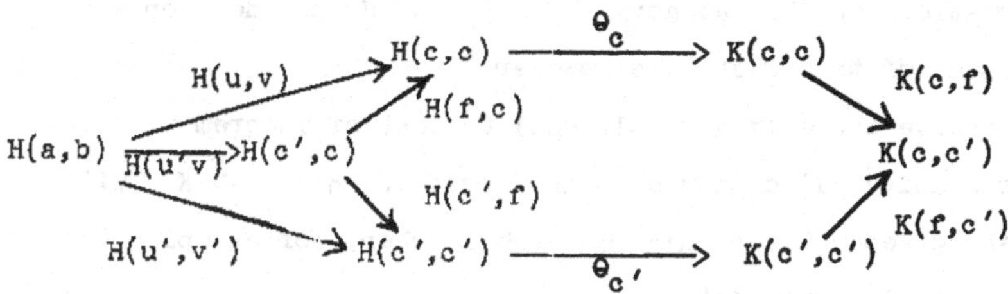

It follows that there exist a unique $\alpha_{a,b} : H(a,b) \longrightarrow \overline{K}(a,b)$ for each $(a,b) \in A^{op} \times A$ such that the square

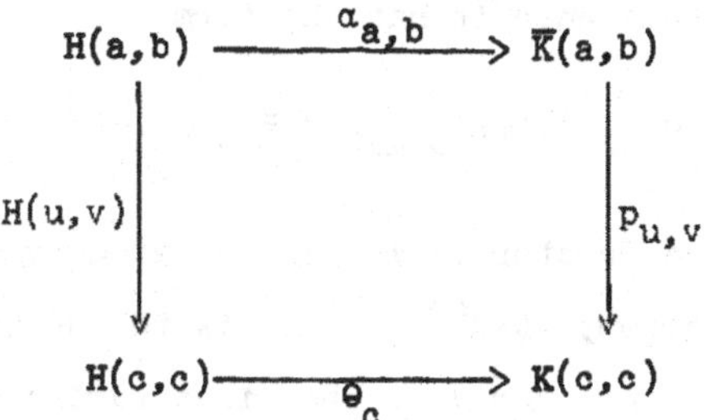

commutes for all commutative diagrams (5) in A. That $\alpha = (\alpha_{a,b}) : H \longrightarrow \overline{K}$ is natural and unique with the property $\Theta = \epsilon\alpha$ is a routine check which we leave to the reader.//

Another way of expressing this result is

(6) $$\text{Dinat } (H,K) \cong \text{Nat } (H,\overline{K}).$$

This equation may also be proved by employing equation (3), a generalized Yoneda Representation Theorem [1] expressed in terms of ends, equation (4) and equation (2) in a straight forward end manipulation.

For A small and B complete there is a natural monad structure on the functor $B^{A^{op} \times A} \longrightarrow B^{A^{op} \times A}$ which takes K to $\bar{K}$. It is generated by the action of the graph of dinatural transformations on the category $B^{A^{op} \times A}$. We do not develop any consequences of this fact here however.

Of course there is a result dual to that of Theorem 2 which provides a universal dinatural transformation $\eta : K \longrightarrow \hat{K}$ with domain any given functor $K : A^{op} \times A \longrightarrow B$ when, for example, A is small and B is cocomplete.

Examples. (i) We will say a functor $H : A^{op} \times A \longrightarrow B$ is dummy in the first variable when it has the form

$$A^{op} \times A \xrightarrow{t \times H'} \mathbb{1} \times B \xrightarrow{\cong} B$$

for $H' : A \longrightarrow B$ a functor ($\mathbb{1}$ is the category with one object and the identity arrow; $t : A^{op} \longrightarrow \mathbb{1}$ is the unique functor). If $H, K : A^{op} \times A \longrightarrow B$ are dummy in the first variable, then a dinatural transformation $\Theta : H \longrightarrow K$ is a natural transformation $\Theta : H' \longrightarrow K'$. Equation (4) gives

$$\begin{aligned} \bar{K}(a,b) &= \int_c (A(b,c) \times A(c,a)) \pitchfork K'(c) \\ &\cong \int_c A(b,c) \pitchfork (A(c,a) \pitchfork K'(c)) \\ &\cong A(b,a) \pitchfork K'(b), \end{aligned}$$

which exists whenever B has powers as big as the hom-sets of A.

(ii) A constant functor sends every object into the same object and every arrow into the identity of that same object (that is, the functor factors through $\mathbb{1}$). The two types of extra-ordinary natural transformations [2] are dinatural transformations $\theta : H \longrightarrow K$ where either H or K is a constant functor. Suppose $K : A^{op} \times A \longrightarrow B$ is the constant functor whose constant value is $x \in B$. Then

$$\bar{K}(a,b) = A(b,a) \pitchfork x .$$

So the extraordinary natural transformations from H to x are in one-to-one correspondence with the natural transformations from H to $A(-,-) \pitchfork x$. Dually, the extraordinary natural transformations from x to H are in one-to-one correspondence with the natural transformations from $A(-,-) \otimes x$ to H.

(iii) Suppose $H : A \longrightarrow B$, $K : A^{op} \longrightarrow B$ are functors. Thinking of these as functors $A^{op} \times A \longrightarrow B$ where H is dummy in the first variable and K in the second, we obtain a definition of a dinatural transformation $\theta : H \longrightarrow K$. It is a family of arrows $\theta_a : Ha \longrightarrow Ka$ in B such that, for all arrows $f : a \longrightarrow a'$ in A, the following square commutes.

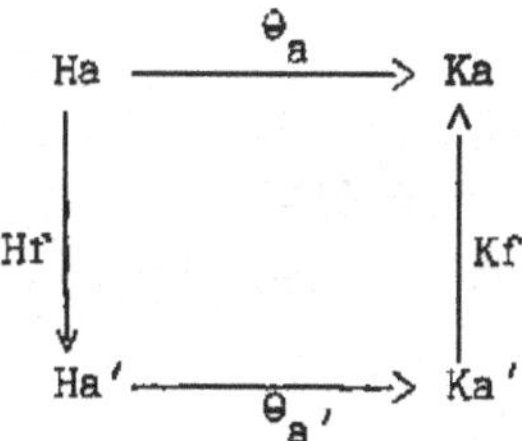

Such things do occur (MacLane). Let V be the category whose objects are vector spaces v (over a field k) and whose arrows $f : v \longrightarrow w$ are linear transformations. Let W be the category whose objects are objects v of V with a linear transformation (inner product) $\langle\ ,\ \rangle : v \otimes_k v \longrightarrow k$ and whose arrows $f : v \longrightarrow w$ are arrows of V satisfying the condition

$$\langle x,y\rangle = \langle fx,fy\rangle \quad \text{for all} \quad x,y \in v .$$

Let $H:W \longrightarrow V$ be the functor which forgets the inner product, let $K:W^{op} \longrightarrow V$ which takes the dual of each vector space, and let $\theta_v : v \longrightarrow V(v,k)$, for $v \in W$, be given by

$$\theta_v(x) = \langle x,-\rangle : v \longrightarrow k .$$

Then $\theta = (\theta_v) : H \longrightarrow K$ is a dinatural transformation.

(iv) Let V be a closed category (if not monoidal, all the better). Let M be the category of multiplicative systems in V; that is, the objects of M are pairs (x,π) where $x \in V$ and $\pi : x \longrightarrow V(x,x)$ is an arrow of V, and the arrows $f:(x,\pi) \longrightarrow (x',\pi')$ are just arrows $f:x \longrightarrow x'$ in V such that the following pentagon commutes.

(7)

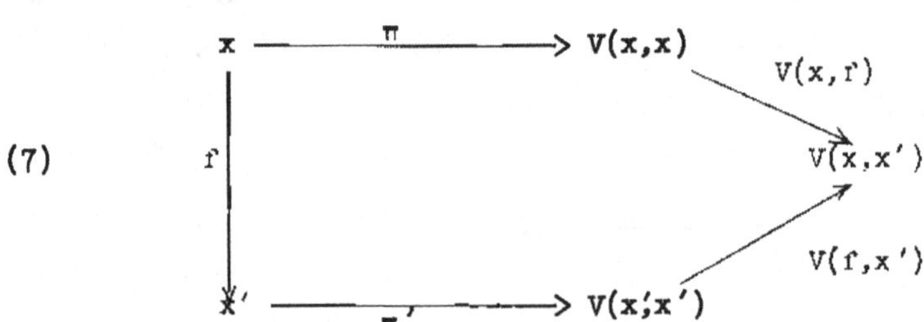

Define $H,K : M^{op} \times M \longrightarrow V$ by

$$H((x,\pi)\ ,\ (x',\pi')) = x'$$
$$H(f,f') = f'$$

$$K((x,\pi)\ ,\ (x',\pi')) = V(x,x')$$
$$K(f,f') = V(f,f') .$$

(If M instead is the category of associative multiplicative systems of V we could take $K = M(-,-)$ provided V has certain limits). Let $\Theta_{(x,\pi)}: H((x,\pi),(x,\pi)) \longrightarrow K((x,\pi),(x,\pi))$ be the arrow $\pi: x \longrightarrow V(x,x)$. The diagram (7) now expresses the dinaturality of $\Theta = (\Theta_{(x,\pi)}): H \longrightarrow K$.

(v) Let A denote any category and let $H: A^{op} \times A \longrightarrow \mathcal{S}$ denote its hom-functor. Define $\sigma_a: H(a,a) \longrightarrow H(a,a)$ by

$$\sigma_a(f) = ff = f^2 \quad \text{for} \quad f: a \longrightarrow a \quad \text{in} \quad A.$$

Then $\sigma = (\sigma_a) : H \longrightarrow H$ is dinatural. This example "squares" the endomorphisms; "taking the n-th power" (for $n \geq 0$) of an endomorphism works equally well (or even negative powers if every endomorphism in A is an automorphism). In fact this example can be elaborated a good deal.

Let $F,G: A \longrightarrow B$ be functors. A dinatural transformation $\Theta: A(-,-) \longrightarrow B(F-,G-)$ will be called a <u>lax natural transformation</u> <u>from</u> F <u>to</u> G. By setting $\varphi_u = \Theta_a(u)$ for $u: a \longrightarrow a$ in A, the definition can be expressed as follows. A lax natural transformation $\varphi: F \longrightarrow G$ is a family of arrows $\varphi_u: Fa \longrightarrow Ga$ in B where $u: a \longrightarrow a$ in A, such that, whenever $v: a \longrightarrow b$, $w: b \longrightarrow a$ are arrows of A, the following <u>square</u> commutes.

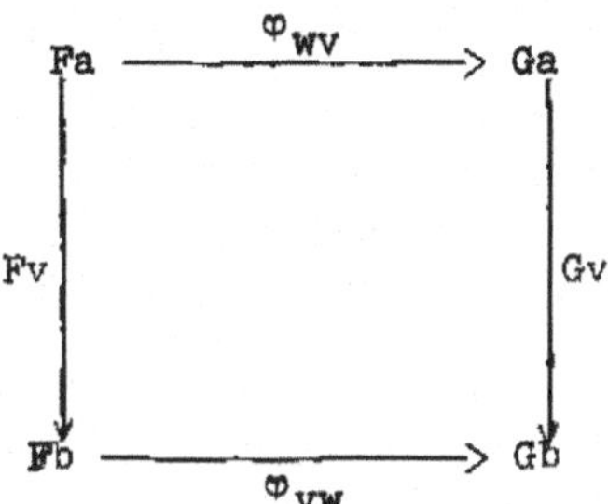

Lax natural transformations compose; there is a corresponding notion of lax functor category and so, a lax closed structure on the category of categories. There are more lax natural transformations than natural transformations in general. For suppose $\alpha : F \longrightarrow G$ is a natural transformation and $m,n \geq 0$ are integers. For each $u : a \longrightarrow$ in A, let $\varphi_u : Fa \longrightarrow Ga$ be given by

$$\varphi_u = (Gu)^n \cdot \alpha_a \cdot (Fu)^m .$$

Then $\varphi = (\varphi_u) : F \longrightarrow G$ is a lax natural transformation.

REFERENCES

[1] B. J. Day and G. M. Kelly, Enriched functor categories. Reports of the Midwest Category Seminar III (Lecture Notes in Mathematics, Springer-Verlag 106, 1969), 178-191.

[2] S. Eilenberg and G. M. Kelly, A generalization of the functorial calculus. J. Algebra 3 (1966) 366-375.

TULANE UNIVERSITY
NEW ORLEANS, LOUISIANA
70118

CATEGORICALLY, THE FINAL EXAMINATION

FOR THE

SUMMER INSTITUTE AT BOWDOIN COLLEGE (Maine) 1969

'I thought I saw a garden door that opened with a key,
I looked again and found it was a Double Rule of Three,
And all its mysteries, I said, are plain as day to me.'

(Verse by the true founder of
Category Theory)

Important Instruction: This is a take-home exam:
Do not bring it back!

Answer as many as possible at a time.

1. Are foundations necessary? To put it another way, given a chance, wouldn't Mathematics float?
2. Describe the category of foundations. Is this a concrete category? A re-enforced concrete category?
3. Discuss the relations and limitations of the foundations set forth by:
   a) Frege-Russell
   b) Bernays-Gödel
   c) Playtex.
4. (Mac Lane's Theorem) Prove that every diagram commutes.
5. Considering a left-adjoint as male and a right adjoint as female, give the correct term for a contravariant functor self-adjoint on the right.
6. Considering a left-adjoint as husband and a right-adjoint as wife, give a precise definition of "marital relations". Do the same for the pre-adjoint situation.

7. Discuss the Freudian significance of exact sequences. (Hint: Consider the fulfillment by one arrow of the kernel of the next.)
8. Find <u>two</u> <u>new</u> errors in Freyd's "Abelian Categories".
9. Trace the origin of the Monads-Triads-Triples controversy to the important paper of St. Augustine.
10. Using theorems from both Freyd and Mitchell, prove that every reflective category is co-reflective. Dualize.
11. Give your opinion of the following exercises:
    a) Ten pushouts
    b) Twenty laps around an adjoint triangle
    c) Two supernatural transformations.
12. Write out at least one verse of
    a) "Little Arrows"
    b) "Doing What Comes Naturally"
    c) "Hom on the Range"
13. Why is the identity functor on $\underline{2}$ called the "Mother Functor"?
14. Write down the evident diagram, apply the obvious argument, and obtain the usual result. (If you can't do it, you're not looking at it hard enough, or, perhaps, too hard.)

Phreilambud

Offsetdruck: Julius Beltz, Weinheim/Bergstr